ART AND ICONOGRAPHY OF THE BUDDHA IMAGES

By

Dr. Shailendra Kumar Verma

EASTERN BOOK LINKERS

DELHI   : :   INDIA

Publishers :

EASTERN BOOK LINKERS

5825, New Chandrawal, Jawahar Nagar,

Delhi-110007 Ph.: 2520287

First Edition : 1994

ISBN : 81-86339-16-7

Rs. 750.00

Laser typesetting & Printing at :

UNIQUE PRINT MEDIA

Phone : 7252362

DEDICATED

To the sacred memory of my Father Late Ramjeelal Das
who spared no pains in education and who would have
been the happiest person if he could see this work,
which all through his life he encouraged me
to complete it very nicely.

FOREWORD

It is well-known that Buddhism arose initially as a simple moral reaction against Brāhmanism, gradually it developed its own mythology, rituals and pantheon but also made a unique contribution to the development of ancient Indian art and culture.

The nature of early Buddhism manifests that it was basically an ethical and moral religion. The Buddha had consistently deprecated the rendering of any special reverence to himself, made no claim to divine prerogative and ignored, if he did not distinctly repudiate and deny, the presence and influence of the supernatural in human affairs. The Buddha said that man himself is the master of his own destiny, according to Him each man must tread the road and win the goal for himself, none helping or hindering him in achieving emancipation. In His ethics there was no room for adoration or worship to gain merit or to achieve emancipation. However, gradually after the *Mahāparinirvāṇa* of the Buddha a tendency evolved in Buddhism which started to regard the Buddha as more than a human being. With the appearance of the Mahāsaṅghikas, the elements of *Bhakti* become dominant in Buddhism. They made the Buddha supramundane and laid special emphasis on the Buddha-*Bhakti*. In fact the Buddha by his superhuman qualities won over the hearts of His devotees and was recognised as a godhead even during His life time. Devotees longed to see Him all the time and as it was not possible to be before Him, they wanted some representation of His anthropomorphic appearance before them. This gave way to development of *Bhakti*-cult in Buddhism and which finally led to the emergence of anthropomorphic images of the Buddha.

The earliest images of the Buddha appeared simultaneously both in Mathura and Gandhāra schools of Art between 1st century B.C.—A.D. The Features of the images of both the schools reveal that they developed independetly, nevertheless, there were interaction between them. Buddhist Art gradually also developed in other parts of India like Sārnāth, South India, Western Indian caves, Kashmir, Eastern India etc. and icono-plastically most of them were influenced by either Gandhāra or Mathura or both the schools of Art.

Though, a number of works have been done on Buddhist Art and Iconography, but the present one, which is done under my supervision, is different to other ones. A comprehensive study has been done not only on the causes of the absence of the Buddha image in early Buddhist Art, but also on the probable causes of the origin of the Buddha images. The Iconographical development of the Buddha images from its inception to 8th century A.D. have been critically analysed which no one has done all together in a book.

The present book is an excellent contribution in the field of Buddhist iconography and would be useful for both researchers as well as students.

My student Dr. Shailendra Kumar Verma has achieved an arduous task in bringing out this work which will fulfil the long-felt need of scholars and students for such a comprehensive study. I wish him all success in life and scholarly pursuits.

—Dr. Sudha Sengupta
Associate Professor (Retd.)
Department of Buddhist Studies,
University of Delhi, Delhi.

FOREWORD

I have gone through the doctoral dissertation of Sri Shailendra Kumar Verma entitled 'Art and Iconography of the Buddha Images'. It presents an analytical and comprehensive study of the images of the Buddha from the earliest times to the eighth century, highlighting their noteworthy aspects in terms of stylistic development, regional features and iconographic orientation. Dr. Verma has spared no pains to collect all the relevant data and put them in a cogent pattern. He has further given them ample visual support. His presentation of facts and literary style are commendable. And I am confident, when published, Dr. Verma's work will be acclaimed by scholars and lay readers alike.

—**Kalyan Kumar Dasgupta**
Bageswari Professor of Indian Art,
Department of Ancient Indian History and Culture,
University of Calcutta;
Formerly Vice-Chancellor,
University of Kalyani, West Bengal.

ACKNOWLEDGEMENT

First of all, I must express my gratitude to those scholars whose research works have been utilized in this book, though the names of all of them could not be incorporated in bibliography.

I am extremely grateful to my learned teacher and supervisor, Dr. Sudha Sengupta, Associate Professor (Retd.), Department of Buddhist Studies, University of Delhi, Delhi, who has always been a great source of kindness and inspiration for me. I do not think it would have been possible for me to complete this arduous task without her inspiration, encouragement and financial assistance. After the sudden demise of my father it was not possible for me to continue further study, but it was she who provided all sorts of help to complete this research work and bailing me out from the financial crisis. She also allowed me to use her own library which proved very useful for me. I will be ever grateful to her.

I am very much grateful to all the teachers of the Department of Buddhist Studies, specially to Prof. K.K. Mittal, the then Head of the Department, Prof. Mahesh Tiwary and Prof. Sanghasen Singh who were kind enough to encourage and helped me by their kind and wise advices occasionally. My grateful thanks go to Dr. (Mrs.) Swati Ganguly, Research Scientist, Department of Buddhist Studies, who helped me a great deal with her advices, encouragement and affectionate sisterly behaviour.

I am thankful to the staff members of Delhi University library for their co-operation.

I am very much grateful to Prof. A.N. Lahiri, Prof. Bela Lahiri and Prof. K.K. Dasgupta, Bageshwari Professor of Fine Arts, Calcutta University, Calcutta for their encouragement in writing this thesis.

I am also thankful to Mrs. Rita Sen and Mr. U.K. Sen for their encouragement and affection towards me.

I am very much thankful to the staff members of National Museum, New Delhi, specially to Mrs. Anamika Pathak and Mr. S.K. Pathak who helped me in getting copies of photographs from the Museum.

I am also very thankful to the staff members of the photo section of the Archaeological Survey of India, New Delhi who helped me in getting copies of photographs from the Survey.

My thank go to the staff members of the library of Indira Gandhi National Centre for Arts, New Delhi for their co-operation.

My friends have always encouraged and inspired me during my research work for which I am very much thankful to all of them.

I will be failing in my duties if I do not acknowledge the help rendered by my mother Mrs. Kunti Das, my elder brother Mr. Birendra Kumar Das, my brother-in-law Mr. Narendra Kumar Verma and other members of the family in continuing my research work peacefully without the worries of the share of my responsibilities towards the family which, after the untimely demise of my father should have come on me also.

I am very much thankful to Mr. S.L. Malhotra, Proprietor, Eastern Book Linkers, Delhi, and his staff members for their co-operation and in bringing out this book in a short time.

Oct. 1994

—Shailendra Kumar Verma
Ashok Vihar-III
Delhi-110052

CONTENTS

(*xiii*)

LIST OF ABBREVIATIONS

ASI	*Archaeological Survey of India*
ASIAR	*Archaeological Survey of India, Annual Report*
ASWI	*Archaeological Survey of Western India*
DN	*Dīgha Nikāya(PTS)*
DPPN	*Dictionary of Pāli proper names, 2Vols.*
EI	*Epigrphia Indica*
IA	*Indian Antiquary*
IHQ	*Indian Historical Quarterly*
JASB	*Journal of Asiatic Society of Bengal*
JRASBI	*Journal of the Royal Asiatic Society of Great Britain and Ireland, London*
MN	*Majjhima Nikāya(PTS)*

LIST OF ILLUSTRATIONS

TABLE OF TRANSLITERATION

अ	आ	इ	ई	उ	ऊ	ए	ऐ
a	ā	i	ī	u	ū	e	ai

ओ	औ	अं	अः	ऋ	क्	ख्	ग्
o	au	ṁ	ḥ	ṛ or ṛi	k	kh	g

घ्	ङ्	च्	छ	ज्	झ्	ञ
gh	ṅ	c or ch	ch or chh	j	jh	ñ

ट्	ठ्	ड्	द्	ण्	त्	थ्	द्	ध्	न्
ṭ	ṭh	ḍ	ḍh	ṇ	t	th	d	dh	n

प्	फ्	ब्	भ्	म्	य्	र्	ल्	व्	श्
p	ph	b	bh	m	y	r	l	v	ś or sh

ष्	स्	ह	क्ष्	त्र	ज्ञ
ṣ	s	h	kṣ or ksh	tr	jñ

INTRODUCTION

The present work, Art and Iconography of the Buddha images is the revised version of the Ph.D. thesis entitled, "Emergence and evolution of the Buddha image (from its inception to 8th century A.D.)" in which the Degree was awarded to the author by the University of Delhi in 1993. The subject is so heart-catching that it is being studied for more than a century last and the very taste allured the present scholar also with insatiable desire to continue it further. The main object of the work is to make a comprehensive study of the factors contributing to the emergence of the anthropomorphic images of the Buddha and their iconographical analysis from its inception to 8th century A.D.

Though, much work had been done on early Buddhist art by many eminent scholars, like, A. Foucher, A.K. Coomaraswamy, Niharranjan Ray, V.S. Agrawala, Stella Kramrisch etc., these works had been done either when the archaeological dating methods were not as accurate as today or they dealt with the general study of the images of the Buddha. Even in some of the works proper credence have not been given to literary sources. Moreover, archaeological excavations have brought to light a number of Buddha images in recent past. So, there is scope to make a fresh and comprehensive study of the Art and Iconography of the Buddha images.

In the earliest period there was no iconographic rules as such for the carving of the Buddha image. The early images of the Buddha which have been discovered from Mathura and Gandhāra are distinct in their styles. In Mathura, the early standing images of the Buddha/Bodhisattva are mainly prototypes of Yakṣa images, and in Gandhāra, the images of the Buddha reflect, *inter-alia*, the Greek 'classical' ideals. However, it seems that there was some short of rudimentary from of iconographic rules prescribed for the carving of the images of great personalities. The *Lakkhana* and the *Mahāpadāna suttanta's* of the *Dīghanikāya* enumerate the thirty-two major and eighty smaller signs of a greatman which is better known as *Mahāpuruṣa lakṣana*. The Buddha is considered as an ideal greatman who possessed as many as thirty-two auspicious physical marks. These signs of greatness include a top-knot on the head i.e. *Uṣniṣa*, a tuft of fine hair between the eye-brows i.e. *Ūrṇā*, long arms reaching up to the knees i.e. *Ājānu-lambita-bāhu*, and webbed fingers of hands and feet i.e. *Jālānguli-pāṇi-pada*. The sculptors of Mathura and Gandhāra who are supposed to have carved the anthropomorphic images of the Buddha for the first time appear to have conformed to this ideal of a greatman. Though, all these signs were not applied to each of the images of the Buddha. However, we may say that the *Mahāpuruṣa lakṣana* was the nucleus of Buddhist iconographic rules.

The thesis consists of nine chapters including, of course, the necessary Introduction.

Chapter I deals with the causes of the emergence of the Buddha image. In this chapter at first we have discussed about the problem of the study as it is well known that scholars are not unanimous regarding the causes of the emergence of the Buddha images. In this connection different theories of many scholars have been reviewed. This very chapter is divided into six sections.

Section I deals with the early Buddhist art tradition and the early images of the Buddha in which mainly we have discussed about the aniconic and the earliest anthropomorphic representation of the Buddha. It is well known that the Buddha was represented in early Buddhist art i.e. before the first century B.C. by different symbols viz, elephants, 'Bodhi-tree', *Dharma-cakra*, *Stūpas*, vacant seat etc. The free standing Aśokan pillars surmounted by different symbols, viz. Lion, Wheel etc. are supposed to be mythologically related to the Buddha. The symbolical representations of the Buddha are also depicted on the gateways and railings of the *Stūpas* from Bharhut, Sānchi, Nāgārjunakoṇḍa, Amarāvati etc.

The anthropomorphic image of the Buddha was evolved in circa first century B.C.-A.D. The coins of Kaniṣka contain the first datable image of the Buddha with Greek legend BODDO,

SAKAMANO BOUDO etc. i.e. the Buddha and Sākyamuni Buddha. Though there are definite evidences that the images of the Buddha were made even before the time of Kaniṣka.

Section II deals with the literary sources regarding the early images of the Buddha. The early Buddhist canonical literature does not mention either in favour of the making of the Buddha image or against it. However, there are some stray references regarding the Buddha image in the *Jātaka* and in some commentarial works, but these are most probably of later origin. *The Mahāvaṁsa* mentions that the images of the Buddha were made during the reign of Aśoka while the *Divyāvadāna* says that the first image of the Buddha was made even during the life of the Master, but in painting.

Section III deals with the Chinese travellers' account regarding the early images of the Buddha. The Chinese travellers', Fa-Hien and Huien-Tsang, who came to India in fifth and seventh century A.D. respectively, mention that the Buddha images were made during His life time.

Section IV deals with the Tibetan and related sources regarding the early images of the Buddha. Tibetan historian Tāranāth claims that the image of the Buddha was made just after His demise. He also mentions that images of the Buddha were also known during the Aśokan period.

But, the above mentioned sources are not corroborated with the archaeological evidences as it is known that before the Kuṣāṇa period there is no evidence available for the anthropomorphic image of the Buddha.

Section V deals with the factors responsible for the absence of the Buddha image in early Buddhist art. As stated above, the Buddha was represented in early Buddhist art i.e. before the first century B.C. through different symbols only. It seems that due to the canonical interdiction, the images of the Buddha were not executed in the early period. Though there is no direct reference regarding the canonical interdiction for the making of the Buddha image in Pāli *Tripiṭaka*, there is an indirect reference in Sarvāstivada *Vinaya*, where Anāthapiṇḍaka prays to the Buddha that if the images of yours are not allowed to be made, please permit at least for the making of the Bodhisattva images and the Buddha then grants the request.

Section VI deals with the causes of the emergence of the Buddha image. This section is further divided into two groups: A and B.

The group A deals with the ideological development in Buddhism in the purview of the emergence of the Buddha image. It is further subdivided into three parts.

Part I deals with the conception of the Buddha in different schools of Buddhism. The Theravādins considered the Buddha as a human being, though, He possessed certain superhuman qualities whereas Mahāsaṅghikas considered the Buddha as *Lokottara* (Supramundane). Like Theravādins, the Sarvāstivādins conceived the Buddha as a human being, but unlike Theravādins they attribute to the Buddha divine, sometimes superdivine powers.

Part II deals with the concept of *Kāya*. The concept of *Kāya* or *Trikāya* i.e. three-fold body of the Buddha contributed to the deification of the Buddha which gave impetus to the emergence of anthropomorphic images of the Buddha.

Part III deals with the evolution of *Bhakti*-cult in Buddhism. The *Bhakti* or devotion was present in Buddhism from the earliest period but not as a cult. With the evolution of *Trikāya* concept and the deification of the Buddha, the Buddha became an object of adoration and worship and gradually in this way devotion evolved as a cult in Buddhism.

The group B deals with the survey of the factors contributig to the emergence of the Buddha image. The *Trikaya* theory, deification of the Buddha, and emphasis on Buddha-*Bhakti* were the prime factors for the emergence of the Buddha image. However, the economic prosperity of the period equally contributed to this task, as, no school of art can flourish or survive without economic support.

Section VII deals with the priority of place where the first image of the Buddha/Bodhisattva evolved - Mathura, Gandhāra or Udyāna-Kashmir. The problem, where the first image of the Buddha was evolved has not been solved unanimously, mainly due to the lack of adequate materials. However, it seems that the images of the Buddha appeared both in Mathura and Gandhāra simultaneously between first centuries B.C. and A.D.

Chapter II deals with the Iconography of the Buddha images in Mathura school of art. The chapter is further divided into four parts.

Part I deals with the historical perspective of the place. Mathura, which is situated on the right bank of the river Yamuna at a distance of 58 km. to the North-west of Agra, was an important centre of trade, commerce and art. During the reign of the Kuṣāṇas, Mathura had established its position as a famous centre of art. After the fall of Kuṣāṇa powers, Mathura gradually lost its position.

Part II deals with Buddhism in Mathura through the ages. Mathura was associated with Buddhism from the life time of the Buddha down to the Imperial Guptas. The Huṇa invasion (5th-6th century A.D.) severely damaged the Buddhist establishments in Mathura.

Part III deals with the Mathura school of Buddhist art. The archaeological finds suggest that Mathura school of art began sometime in the 3rd or 2nd century B.C. and the early art products of this school are related with Bharhut and Sānchi school. This school established its own style in Indian art.

Part IV deals with the Iconography of the Buddha images. The early standing images of the Buddha from Mathura school are mainly the prototypes of Yakṣa images. During the Kaniṣka period the images were not uniform, they differed from image to image. The images of the Buddha of the Huviṣka period show penetration and culmination of Gandhāran traits in Mathura school whereas the images of the Gupta period became slim and slender and the expression reflected the inner bliss and serenity, may be due to the impact of Yogāchara school of Buddhist thought.

Chapter III deals with the Iconography of the Buddha images in Gandhāra school of art. The chapter is further divided into four sections.

Section I deals with the historical perspective of the region. Gandhāra is situated in North-western frontiers of undivided India.

In the 6th-5th centuries B.C. Gandhāra was the part of Achaemenid empire of Persia and for a short period in 4th century B.C. it was occupied by the armies of Alexander and later on it was conquered by Chandragupta Maurya. After the fall of the Mauryan empire Gandhāra was occupied by Scythians or Śakas, Pārthians and Kuṣāṇas respectively who included it into their empire. In about 5th century A.D. the Huṇas captured Gandhāra and ruled over it up to the early 6th century A.D. when Hindushahiyyas dethroned them. The Hindu-shahiyyas ruled over Gandhāra untill the advent of Sultan Mohmud of Ghazni in 11th century A.D.

Section II deals with Buddhism in Gandhāra through the ages. It seems that Buddhism was systematically introduced into the Gandhāran region by *Thera* Majjhantika after the convention of the third Buddhist Council. Huien-Tsang mentions that he had seen a few *Stūpas* erected by Aśoka near Pushkalāvati. In this region both Hīnayāna and Mahāyāna tradition of Buddhism was prevalent. The Sarvāstivāda school was more popular here. Huien-Tsang further mentions that of the *Stūpas* and monasteries he saw, some of them were in dilapidated condition.

Section III deals with the Gandhāra school of Buddhist art. The Gandhāra school of art has a distinct place in the history of Buddhist art. This school of art has been called by different names viz., Indo-Bactria, Indo-Greek, Graeco-Buddhist and Indo- Roman, mainly due to their influences on Gandhāran art. The Gandhāra school of art most probably flourished in first century B.C. and lasted till the Huṇa invasion in 5th century A.D. The art products of this school are mainly Buddhist in nature.

Section IV deals with the Iconography of the Buddha images. The images of the Buddha discovered at Butkara I (Pakistan) belonging to the period first century B.C. - A.D., is considered to be the earliest representation of the Master in Gandhāran region. The general features of the Gandhāran Buddha images are: seated or standing images are conceived as a short, rather stocky, and the position of the body is invariably frontal whenever He appears independently. The hair is arranged in waves. The eyes are usually open and there is *Ūrṇā* between the eyebrows. He usually wears moustache. The body is covered in a heavy monastic cloak that hangs in deep folds and covers both the shoulders. Here the Buddha is usually seated on lotus throne.

Chapter IV deals with the Iconography of the Buddha images in Sārnāth school of art. The chapter is further divided into four parts.

Part I deals with the historical perspective of the place. During the age of the sixteen *Mahājanapadas*, Sārnāth was related to Kāshi. During the period of Mauryan empire it was obviously part of it. Though Sārnāth is famous as a place where the Buddha had delivered his first sermon, it came into prominence during the Gupta period mainly as a centre of art.

Part II deals with Buddhism in Sārnāth through the ages. As stated above, Sārnāth is the birth place of Buddhism, here the Buddha had turned the wheel of law for the first time. The remnants of Aśokan *Stūpa* at Sārnāth show the prevalence of Buddhism in the place. The Dhamekh *Stūpa* which was erected during the Gupta period is a witness about the prevalence of Buddhism at the time. However, Buddhism lingered at Sārnāth even after the Guptas.

Part III deals with the Sārnāth school of Buddhist art. The Sārnāth school of art came into prominence during the Gupta period, though it might have its origin beforehand. The early images of Sārnāth school are mainly based on the ideals of Mathura school. However, gradually it established its own ideals.

Part IV deals with the Iconography of the Buddha images. The images of the Buddha appeared at Sārnāth school during the Kaniṣka's period, but those images are supposed to be imported from Mathura school. During the Gupta period we find a break-through in the ideals of Sārnāth school of art, numerous Buddhist images which are exquisitely beautiful, came into existence. The general features of the images of the Buddha from Sārnāth school are: the faces of the images are oval and the heads are covered with short curly hairs. The draperies are foldless and reduced in volume. In the later period the halo is replaced by a large almond shaped back piece which is larger than the image.

Chapter V deals with the Iconography of the Buddha images in Kashmir school of art. The chapter is further divided into four sections.

Section I deals with the historical perspective of the place. The ancient Kingdom of Kashmir was part of the present State of Jammu and Kashmir. The political history of Kashmir before the advent of Aśoka is not very much clear. Kalhaṇa says that Aśoka had founded the city Srinagar and built a number of *Stūpas* in that area. After the fall of the Mauryan empire, the Indo-Greeks, the Śākas, the Pārthians and the Kuṣāṇas gradually extended their sway over most part of the North-western regions of India, so, Kashmir naturally came under their powerful leadership. After the fall of the Kuṣāṇa powers, Kashmir came into the hands of local Kings. The Huṇas captured Kashmir in the fifth century A.D. The Huṇas were followed by Kārkoṭas. The most famous Kārkoṭa king was Lalitāditya Muktāpiḍa (724-761 A.D.).

Section II deals with Buddhism in Kashmir through the ages. Buddhism was most probably introduced into Kashmir during the Aśokan period, if not earlier. After the disintegration of the Mauryan empire, the Indo-Greeks, and the Kuṣāṇas etc. who ruled over Kashmir, extended their support to Buddhism. It is well known that the fourth Buddhist council was held at Kashmir during the reign of Kaniṣka. Kashmir was one of the centres of Sarvāstivādins.

Section III deals with the Kashmir school of Buddhist art. Kashmir school of art yielded a number of Buddhist images in early medieval India. The features of the images belonging to this school reflect the influence of both Gandhāra and Mathura schools.

Section IV deals with the Iconography of the Buddha images. As stated above, the Buddhist images of Kashmir school were very much influenced by both Gandhāra and Mathura schools, it differs very little from each other. Here we find some ornamented images of the Buddha.

Chapter VI deals with the Iconography of the Buddha images in western Indian caves. The chapter is further divided into four parts.

The part I deals with the historical perspective of the region. Western India was also known as Aparānta. During the period of the *Mahājanapadas* Avanti was an important kingdom of western India. The Mauryas, the Sātavāhanas, the Vākāṭakas and the Chālukyas of Badāmi respectively ruled over Western India upto the 8th century A.D.

Part II deals with Buddhism in Western India. Buddhism was systematically introduced in Western India during the reign of Aśoka by Dhammarakṣita a Yoṇaka or Ionian Greek. Buddhism lingered in Western India upto the 8th century A.D.

Part III deals with the Western Indian Buddhist caves and sculptures. The caves of Western India have broadly been divided into two sections - Hīnayāna and Mahāyāna. The caves of Hīnayānic phase contain the *Stūpa* in the sanctum sanctorum whereas Mahāyānic caves are those which contain the images of the Buddha etc. The caves of Kārle, Nāsik, Kanheri, Bāgh, Ellora and Ajantā contain the images of the Buddha.

Part IV deals with the Iconography of the Buddha images. The images of the Buddha from the Western Indian caves are generally influenced by Sārnāth school. The bodies of the seated Buddha images are heavy and wears foldless drapery. Here the images of the Buddha are also in relief.

Chapter VII deals with the Iconography of the Buddha images in Andhra school of art. The chapter is further divided into four sections.

Section I deals with the historical perspective of the region. Before the rise of the Mauryas the history of South India is not very much clear. However, during the reign of the Mauryas, some portions of South India came under their domain and the rest of the territory were divided into three kingdoms: 1. The Chola, 2. The Chera, and 3. The Pāṇḍya. The Mauryas were followed by the Sātavāhanas in South India, and after the fall of the Sātavāhana power, South India was divided into different kingdoms, among them, the Ikṣavākus established their sway over the Andhra region. The Ikṣavākus were supplanted by the Pallavas who ruled upto 8th century A.D. Apart from these there were a few dynasties who ruled over different parts of South India which have been discussed in the main body of the thesis.

Section II deals with Buddhism in South India through the ages. Scholars are not unanimous when exactly Buddhism was introduced into South India, however, Amarāvati finds support that Buddhism was known in Andhradeśa by fourth century B.C. i.e. in the pre-Aśokan period. The Mahāsaṅghikas had strong hold in Andhra. The Mahāyāna Buddhism also held the position from the very time of its appearance. Buddhism lingered in some parts of South India upto the 14th century A.D.

Section III deals with the Andhra school of Buddhist art. The history of Andhra school of art can be traced back from the period 2nd-1st century B.C. The subject matter of the early phase of Buddhist art of South India are symbolic in character and narrative in nature as in the case of Bharhut, Sānchi etc. The anthropomorphic images of the Buddha appeared in South India in 2nd and third century A.D. The images are both in relief and in round which are unearthed in Amarāvati, Jaggayyapeṭṭā, Bhaṭṭiprolu, Nāgārjunakoṇḍa etc.

Section IV deals with the iconography of the Buddha Images. The credit for the introduction of the Buddha images in Andhra school is generally given to the Chaityaka sect, an off-shoot of the Mahāsaṅghikas. The early images of the Buddha from Amarāvati show certain heaviness like Mathura school. Hellenism also influenced the Andhra school. However, gradually Andhra school established its own ideal or style.

Chapter VIII deals with the Iconography of the Buddha images in Eastern school of art. The chapter is further divided into four sections.

Section I deals with the historical perspective of the region. We have confined our area of studies into these three states of Eastern India viz, Bihar, Bengal and Orissa only.

A. *BIHAR* : Two of the ancient *Mahājanapadas*, Aṅga and Magadha and the adjoining areas later on came to be known as Bihar, because of the predominence of Buddhist monasteries (*Vihāras*) which the area was dotted with. Gradually Magadha became more powerful.

During the period of the Mauryan empire Bihar became more prominent. The Mauryan rulers had their capital at Pāṭaliputra (modern Patna). The Mauryans were followed by the Śuṅgas and the Kāṇvās respectively. The Guptas came into power in early 4th century A.D. After the fall of the Gupta empire in 6th century A.D., several new dynasties came into power. The Pālas came into power in early 8th century A.D.

B. *BENGAL* : In modern times Bengal, which is situated between the rivers Ganga and Brahmaputra, means Bengali speaking areas of independent India and Bangladesh. In early period it was divided into a number of small kingdoms viz, Gauḍa, Vaṅga, Samataṭa, Harikela, Puṇḍravardhana, Tāmralipta etc. However, in course of time Gauḍa and Vaṅga became more popular and the remaining states were merged with either of them.

The early history of Bengal is not very much clear, however, on the basis of the Mahāsthāna inscription we may surmise that Bengal was the part of the Mauryan empire and after the fall of the Mauryan empire, Bengal was ruled by many dynasties. One of the famous kings of Bengal was Śaśāṅka whose period was seventh century A.D. The Pālas came into power in the 8th century A.D.

C. *ORISSA* : Orissa which is situated in the Eastern India and south of the river Mahānadi, was an important centre of Buddhism in early medieval India. The early history of Orissa is not very much clear. Perhaps first it came into prominence during the Kaliṅga war. Khāravela was an important king of Orissa whose period was first century B.C. The reign of the Bhaumakaras is considered to be the glorious age in the history of Orissa. The Bhaumakaras were contemporary to the Pālas of Bengal and they ruled over Orissa upto about 10th century A.D.

Section II deals with Buddhism in Eastern India through the ages.

A. *BIHAR* : Bihar is described as the sanctum sanctorum of early Buddhism. Though traditionally the Buddha had delivered his first sermon at Ṛsipatana (Sārnāth), but elements of Buddhism might have been preached for the first time in Bihar (Bodhgayā) by the Buddha himself to Tapassu and Bhallika who were on their way to

Bodhgayā where they met the Buddha just after his enlightenment and became His converts. Buddhism lingered in Bihar till the 14th century A.D.

B. *BENGAL* : The early history of Buddhism in Bengal is obscure. However, it is said that Buddhism might have appeared in Bengal at least in the early years of the christian era. Buddhism became very much popular during the Pāla period, because the rulers were staunch followers of Buddhism. This is the period when Buddhism and Brāhmanism were coming closer to each other.

C. *ORISSA* : Generally it is said that Buddhism was introduced into Orissa by the two merchants, Tapassu and Bhallika, but this is not corroborated with the historical facts. It seems that Buddhism might have systematically introduced into Orissa during the Aśokan period. The reign of the Bhaumakaras in Orissa is considered to be the golden age in the history of Buddhism. These rulers were devout Buddhist as is evidenced from the epithets used by some of the kings as *Paramatathāgata* and *Paramasaugata*.

Section III deals with the Eastern school of Buddhist art.

A. *BIHAR* : The history of Buddhist art in Bihar can be traced back from the days of the Mauryan rule. The free-standing Aśokan pillars surmounted by different animal figures are supposed to be symbolically representing the Buddha. Buddhist images continued to be made in Bihar at least up to the Pāla period.

B. *BENGAL* : The history of Buddhist art in Bengal began with the Gupta rule if not earlier. The motifs of art mainly exhibit the well-known characteristics of Sārnāth school, more or less combined with the emotionalism of its eastern version. A number of Buddhist images were made during the Pāla period in Bengal, most of them being carved on the stele. The Bengal school of art continued to flourish atleast upto circa 1200 A.D.

C. *ORISSA* : Orissa yielded a number of Buddhist images in early medieval period. Udayagiri, Ratnagiri, Lalitgiri, Solampur etc. were the important centres of Buddhist art. Buddhist images of Orissa are aesthetically beautiful,though, somewhat influenced by the art of Sārnāth school.

Section IV deals with the Iconography of the Buddha images.

A. *BIHAR* : Buddhist images have been discovered from different parts of Bihar viz, Bodhgayā, Kumrahār, Nālandā, Sultanganj etc. The images are somewhat influenced by both Mathura and Sārnāth schools.

B. *BENGAL* : Bengal has yielded a number of Buddhist sculptures, discovered from different parts of the region. The features of the early images are based on the basis of Sārnāth school. The Biharail Buddha image is considered as the earliest representation of the Master in Bengal; gradually Bengal developed its own peculiarities.

C. *ORISSA* : Though, the images of the Buddha from Orissa are influenced by the ideals of Sārnāth school, they have their own peculiarities also e.g. the images are huge and life-like or even bigger. Though the draperies are almost copies of Sārnāth school, facial expressions are somewhat different.

Chapter IX deals with the conclusion which is incorporated in the body of the thesis.

Due to some unavoidable reasons, all the described images are not incorporated.

CHAPTER - 1

EMERGENCE OF THE BUDDHA IMAGE

1.1   *Early Buddhist art tradition and the early images of the Buddha.*

1.2   *Literary sources regarding the early images of the Buddha.*

1.3   *Chinese travellers' and related accounts regarding the early images of the Buddha.*

1.4   *Tibetan and related sources regarding the early images of the Buddha.*

1.5   *Factors responsible for the absence of the Buddha images in early Buddhist art.*

1.6   *Causes of the emergence of the Buddha image.*

1.6.   *A. A study of the ideological development in Buddhism in the light of the emergence of the Buddha image.*

        Part- I. Conception of the Buddha in different schools of Buddhism.

        Part- II. The concept of Kāya.

        Part- III. Evolution of Bhakti-cult in Buddhism.

1.6.B. *A survey of the factors contributing to the emergence of the Buddha image.*

1.7   *Priority of places where the first images of the Buddha/Bodhisattva evolved-Mathura, Gandhāra or Udyāna- Kashmir?*

1. EMERGENCE OF THE BUDDHA IMAGE

The problem of the emergence of the Buddha image has not been unanimously solved. Though much work have been done on early Buddhist art, still scholars are not unanimous in their opinions about when, where and how the images of the Buddha evolved. However, it seems that combination of such factors as the Brāhmanic *Bhakti* movemment, the Sarvāstivādins' and the Mahāsaṅghikas' inspiration, material prosperity of the period and the patronage of kings etc. gave impetus to the emergence of the Buddha image. Subjectivism of scholars opting for either Greek primacy or Indian genius has uptill now influenced their arguments on the problem of fixing up the place and date of the emergence of the Buddha image. In this respect scholars are divided into three camps : Gandhāra, 2. Mathura, and 3. Udyāna-Kashmir. The theory of Gandhāra origin of the Buddha image is led by Foucher,[1] Grunwedel,[2] Smith,[3] Tarn, [4] etc. The theory of Mathura origin of the Buddha image is advocated by Iwaski Masunavi, Goloubew,[5] Coomaraswamy,[6] Lohuizen,[7] Stella Kramrisch,[8] V.S.Agrawala,[9] and others and the theory of Udyāna-Kashmir origin of the Buddha image is put forth by A.K.Narain.[10] Some scholars like K.K.Dasgupta and B.N.Mukherjee opined that both Gandhāra and Mathura schools of art developed independently and perhaps the images of the Buddha evolved in both schools simultaneously.[11] The arguments of the supporters of Gandhāra origin of the Buddha image are that image-making was foreign to the Indian mind and it were the 'more civilized' Greeks or Romans, settled in the North-Western province of India known as Gandhāra, who were responsible for the making of the first anthropomorphic image of the Buddha, otherwise, the Indians were using only aniconic symbols to represent the Buddha. But this theory is refuted by many scholars

---

1.   A. Foucher,*Beginning of Buddhist Art and other Essays*, London,1914,pp.iii ff.

2.   A. Grunwedel, *Buddhist Art in India*, Delhi,1972, pp.162ff.

3.   V.A.Smith, *A History of fine Art in India and Ceylon*, Oxford,1870.pp.51-52.

4.   W.W.Tarn *Greeks in Bactria and India*, Cambridge, 1938,p.396.

5.   *Victor Goloubew, Quoted in " The Image of the Buddha"*, by D.L.Snellgrove (Ed.) Delhi,1978,p.59.fn.15

6.   A.K.Coomaraswamy, *History of Indian and Indonesian Art*,New Delhi,1972,pp.59-60

7.   J.E.Van Lohuizen-de Leeuw, *The Scythian Period* Leiden,1949,p.154ff.

8.   Stella Kramrisch, *Indian Sculpture*, Delhi,1981, pp. 39-40.

9.   V.S.Agrawala, *Bhārtiya Kalā*, Varanasi,1966,pp. 285-294.

10.   A.K.Narain, "First Image of the Buddha and Bodhisattva" in *Studies in Buddhist Art of South Asia*, New Delhi,1985,p.2

11.   K.K.Dasgupta,"Origin of the Buddha Image" in *Studies in Ancient Indian History* (D.C. Sircar Commemoration Volume), Delhi, 1988, pp.151-152; B.N. Mukherjee, "Forward" in R.C. Sharma, *Buddhist art of Mathura*, Delhi, 1985, p.x.

on the ground that image making was already in vogue and prevalent in India atleast before the emergence of the Buddha image and it was due to the indigenous socio-religious needs that the first image of the Buddha was made at Mathura, icono-plastically based on earlier art tradition rather than any extraneous art tradition whatsoever. This long drawn debate has been further complicated by the lack of mutual consent on the absolute date of Kaniṣka.[12] Nature and limitation of archaeological and numismatic evidence and their inadequate correlation with the literary and epigraphic documentation also has added to this probelm. Frequent change of rules, dynasties and eras used by them before and at the beginning of the christian century both at Gandhāra and Mathura has also posed a problem. Nature, influence, date and development of Gandhāran art have further added to this complication. In spite of this, scholars have tried to solve this problem time and again.

1.1.   Early Buddhist art tradition and the early images of the Buddha.

It is suggested that perhaps Buddhist iconography was articulated for the first time in the art of Aśoka through his free-standing pillars found at different places with wheel and animal capitals, are supposed to be mythologically related to the Buddha.[13] But here one should be clear that the symbols exhibited in Aśokan art had equal place both in Buddhism and Brāhmanism. Since Aśoka had done a lot for Buddhism, so we may say that these symbols were related to the Buddha. Monuments of early Buddhist art found at Bharhut, Sānchi, Bodhagayā and Amarāvati from second century B.C. onwards are aniconic which illustrate the different events of the life of the Buddha. Anthropomorphic images of the Buddha are also depicted but only for His previous births (*Jātaka*).

On the basis of archaeological and numismatic evidences we may say that the anthropomorphic images of the Buddha appeared both in Gandhāra and Mathura in first century B.C. - A.D. in response

---

12.   A.L. Basham (Ed.), *Papers on the date of Kanishka*, Leiden, 1968.
13.   K.K. Dasgupta, "Aśokan Art - Why and how for Buddhist", *Proceedings of Indian History Congress*, XXXI, 1969, pp.56-60.

to the popular impulse in Buddhist community. Representation of the cross-legged seated figure on a class of coins of Śaka King Maues (circa 95-75 B.C.) is considered to be the earliest anthropomorphic depiction of the Buddha and on the basis of this figure A.K. Narain evolved a theory of Udāyana- Kashmir origin of the Buddha/Bodhisattva image since Maues sometime ruled from that area and he was the patron of Buddism.[14] But scholars are not unanimous to accept it as the first anthropomorphic image of the Buddha. Further discussions on this topic are being made in the last section of this chapter. The first datable image of the Buddha appears on the coins of Kaniṣka with Greek character BODDO, SAKAMANO BOUDO and METRAGO BOUDO,[15] indicating the Buddha, Sākyamuni Buddha, Maitreya Buddha. A Kharoṣṭhi Inscription discovered from the Swāt Valley mentions that anthropomorphic images of the Buddha were made even before the Kaniṣka period.[16] A number of sculptured panels which display *inter alia* the images of the Buddha in a group of monuments discovered at Butkara I (Pakistan) which have been assigned to the period first century B.C. - A.D. [17]mainly on numismatic grounds, are considered to be the earliest anthropomorphic representation of the Master from the Gandhāran school.

It is found that in Mathura also the images of the Buddha were being executed before the time of Kaniṣka but they were named as Bodhisattva images, may be due to the age old injunction forbidding the representation of the Buddha in human form. Such practice was continuing even during and after the Kaniṣka period. [18]

---

14. A.K. Narain,*Op. Cit.*, p.16; A.K. Coomaraswamy,*The Origin of the Buddha Image*, Delhi, 1972, fig. 6-12.

15. Joe Cribb, "Re-examination of the Buddha images on the coins of Kaniṣka ... in A.K. Narain (ed.), *Studies in Buddhist Art of South Asia*, Delhi, 1985, p.63; A.K. Coomaraswamy, *History of Indian and Indonesion Art*, Delhi, 1972, p.59, fig. 123.

16. H.W. Bailey, " A Kharoṣṭri Inscription of Senavarma, King of Oḍi", *JRASBI*, 1980, mp.I, pp.1.ff; for details see p. 169.

17. D. Faccenna, "Excavations of the Italian archaeological Mission (ISMEO) in Pakistan : Some Problems of Gandhara Art and Architecture", *Central Asia in the Kushana Period.* Vol.I, (Dushanbe conference), Moscow, 1975, pp. 150ff.

18. K.K. Dasgupta, *Origin of the Buddha Image ...*, p.151; R.C. Sharma, *Buddhist art of Mathura ...*, pp. 154- 166.

However, later Buddhist tradition says that the anthropomorphic images of the Master were made even during His life time and this is also corroborated by the accounts of the Chinese travellers'. But we find a big gap between the first representation of the Buddha image in textual account and the archaeological and numismatic evidence. Some scholars have been critical of the approach adopted by art historians and have insisted on giving more credence to the textual accounts about the early representation of the Buddha image. Lewis R. Lancaster has been critical of the approach of art historians for laying too much stress on factors like popular deification of the Buddha or the development of the *Trikāya* concept as the origin of the Buddha image.[19] Padmanabh S. Jaini has also criticised the art historians for not thoroughly examining the textual evidence regarding the origin of the Buddha image and not giving more credence to the accounts of Chinese travellers.[20]

1.2. Literary sources regarding the early images of the Buddha.

The Pāli commentaries which were perhaps composed sometime between 5th and 6th century A.D. does not mention about image worship of the Buddha. As for example, in the commentaries on the *Vinaya*, the *Dīghanikāya*, the *Majjhimanikāya* and the *Aṅguttaranikāya* and in the *Vibhaṅga* of the *Abhidhamma Piṭaka*, monks are instructed to go and worship the *Caitya* and *Bodhi-tree*; the Buddha image is not mentioned.[21] In these commentaries where retribution of sinful acts (*Anantariya kamma*) such as breaking of a *Caitya*, cutting of *Bodhi-tree* or damaging relics, are described as crimes, the Buddha image is not mentioned.[22] Even in the commentaries on the Pāli *Abhidhamma* texts viz., the *Aṭṭhasālini* (commentary on *Dhammasaṅgani*) and the *Sammohavinodani* (commentary on *Vibhaṅga*) wherein it is stated that the

---

19. Lewis R. Lancaster, ''An Early Mahayana Sermon About the Body of the Buddha and the making of Image'', *Artibus Asia*, XXXVI, 4, 1974, pp. 287-291.

20. Padmanabh S. Jaini, ''On the Buddha Image'', in A.K. Narain (Ed.), *Studies in Pali and Buddhism* (A Homage Volume to the Memory of Bhikkhu Jagdish Kashyap), Delhi 1979, pp. 183-188.

21. Sri Jñanakirit, ''The Commentaries of Buddhaghosa'', *The Mahabodhi*, July, 1968, p. 216.

22. *Iibd.*

joy or ecstasy derived from looking at the Buddha or contemplating on Him (*Buddhālambana piti*) can be obtained by looking at a *Caitya* or a *Bodhi-tree*, there is no reference to the Buddha-image.

However, in spite of this conspicuous absence of any reference to the Buddha-image in the above mentioned works, there are some stray references, like, one in the *Jātaka* and two others in the commentarial works of the *Majjhima* and the *Aṅguttara Nikāya*. It is stated in the *Kusajātaka* of the *Khuddakanikāya* that golden images of the Buddha were made by the chief Smith of a king.[23] Buddhaghoṣa mentions in his commentary on *Vinay* about the age-old practice of meal offering to the Buddha-image.[24] Wisemen, before taking their food and drink offered the same to the image of the Buddha or relic casket symbolising the Buddha.[25] Some other texts, the *Papañcasudani* and the *Manorathapurani* also states that the image of the Buddha is worthy of adoration if a relic is enshrined in it.[26] The Pāli commentarial sources mention that the image of the Buddha is the *Uddesika Caitya* only, not a concrete representation of the actual form of the Buddha, but translation of some ideas about him into an artistic shape.[27]

Apart from these, some of the post-canonical Buddhist works also refer to the Buddha-image, rather in advanced form. A Sanskrit text of image-making says that the Buddha himself had sanctioned regarding the making of His image but only after His demise, for the purpose of worship and adoration.[28] According to tradition Kashyapa had composed six *Sūtras* for the making of the images of the Buddha and even if a man makes an image as small as a grain of barley, it

---

23. B.N. Chaudhury, "Date of Buddha-worship", *The Mahabodhi*, Vol. 75, No. 7 (July, 1967), p.244.

24. *The Vinaya Aṭṭhakathā*, III, 264-5-9.V.R.F. Gombrich, Precept and Practice, p. 121.

25. *Ibid.*

26. Dhammakitti Sri Dhamma (Ed.), *Papañcasudani*, 1962, part II, p. 22; *Manorathapurani*, (Explanatory Notes) Part II, p.83.

27. Bhikku Silabhadra, "The Buddha Image", *The Mahabodhi* Vol. 61, No.7, July, 1953, p. 271.

28. "*Mayi mrite parinirvrite sati. Pujā satkarartham pratimālakshana*", - Q.V-O.C. Gangoly, "The Buddha Image", *The Mahabodhi*, Vol. 61, No. 5&6, pp. 190-92.

would be conducive to his immeasurable merits.[29] The *Divyāvadāna* says that when king Bimbisāra desired to have a representation of the Buddha, then the Buddha agreed to allow his shadow to fall upon a cloth in order to facilitate the painting of his figure on it. He also desired to fill the outline with colour and some religious sentences.[30] Similar accounts are also recorded in Chinese and Tibetan sources, according to them only *Viśvakaramā*, the divine architect, could succeed in drawing in outline the form of the Buddha.[31] In one of the two chief Sinhalese chronicles, the *Mahāvamsa* mentions that king Parākrambāhu had employed many skilled painters to paint on a cloth surpassing likeness of the Buddha.[32] According to this account during the reign of great Sinhalese kings, images and image-houses were built. [33] An indirect or negative evidence regarding the Buddha-images and image-houses is mentioned throughout this chronicle which contains the accounts of kings presenting gifts to statue-carvers, and rival kings carrying away the images of the Buddha to other territories after conquering them.[34] It is also mentioned in a number of lengthy accounts of this chronicle that the Sinhalese kings of the past and their royal household got the images of the Buddha in stone, gold, bronze etc. and they installed them in image-houses.[35] Where as some Sinhalese kings such as Dappali, Udaya, Parākaramabāhu, Kittisiri Rājasiha, decorated the images of the Buddha with costly ornaments and celebrated festivities in their honour.[36] Some other royal patrons namely Jetthatissa, Mahāsena, Sirimeghavāhana and Dhātusena used to worship the images of the Buddha.[37] It is further stated that the Sinhalese kings namely Silameghavanna and Kittisiri Rajasiha made different kinds of offering

29. J. Takakusu, *A Record of the Buddhist Religion ...by* I.Tsing, Oxford, 1896, pp. 151-52.

30. *Rudrāyanavadāna*, (Divyāvadāna), (Ed.), E.B. Cowell, Cambridge, 1886, p. 466.

31. D.N. Varma, "The Image of Buddha", *The Mahabodhi*, Vol. 77, No. 11-12, 1969, pp. 383-84.

32. L.C. Wijesinha Mudaliya (Tr.), *The Mahāvamsa*, Part II, Ceylon, 1909, p. 241.

33. *ibid.*, pp. 4,14.

34. *Ibid.*, pp. 75,304.

35. *Ibid.*, pp. 29- 30,47,50,93,122,166,213-14,217,219,240,252,267,282,310.

36. *Ibid.*, pp. 30,70,122,162,306-7.

37. George Turner, *The Mahāvamsa* (Tr.), part I, 1837,pp.156- 174.

to the images of the Buddha such as gems, bowls of silver and gold, robes umbrella, various kinds of food, flowers and perfumes.[38] *The Mahāvaṁsa* further mentions that an image of the Buddha was made during the reign of Aśoka and another image of the Buddha was installed in the relic chamber of the *Mahāthūpa* at the time of king *Duṭṭhagāmini(C. 101.77 B.C.).*[39]

The above mentioned literary sources can not be taken as historical facts, first they are not contemporary to the images and these works must have been written long after the images were executed, so we can not put reliance on these texts. Actually they were written down when Buddhist schools of art were already in existence, so we may surmise that these works were written to give a canonical sanction to the image making. Even these sources are not corroborated with the archaeological finds.

However, in spite of non-corroboration of the literary and archaeological sources, we may say that there is a long tradition of image making which we can not ignore completely.

1.3. Chinese travellers' and related accounts regarding the early images of the Buddha.

In this section we will discuss the accounts of the Chinese travellers and some other related sources.

The Chinese travellers Fa-Hien and Huien-Tsang who came to India in the sixth and seventh century A.D. respectively, mentioned in their accounts that the images of the Buddha were being made even during His life time. Fa-Hien's account regarding the image of the Buddha is as follow:

"When the Buddha went up to heaven for ninety days to preach the faith to his mother, king Prasenjit longing to see him, caused to be carved in sandal-wood from the Bull's head mountain an image of the Buddha and placed it where the Buddha usually sat. Later on, when the Buddha returned to the shrine, the image straight away quitted the seat and came forth to receive him, then the Buddha

---

38. See *The Mahāvaṁsa*, part II, pp. 22,304-4.
39. K.K.Dasgupta, *"The origin of the Buddha image"*...,p.41

cried out, 'Return to your seat, after my disappearance you shall be the model for the four classes of those in search of spiritual truth."[40]

Another Chinese traveller, Huien-Tsang, who came to India during the reign of Harṣavardhan of Kanauj, interestingly, gives almost similar account as that of Fa-Hien. His account regarding the image of the Buddha is as follow:

"In the city within an old palace, there is a large *Vihāra* about 60 feet high; in it is the figure of Buddha carved out of sandal-wood, above which is a stone canopy. It is the work of king Udayana. By its spiritual qualities it produced a divine light, which from time to time shines forth. The princes of various countries have used their power to carrry off this statue, but although many men have tried, not all the number could move it. They, therefore, worship copies of it, and they pretend that the likeness is a true one, and this is the original of all such figures.

When the Tathāgata first arrived at complete enlightenment, he ascended up to heaven to preach the law for the benefit for his mother, and for three months remained absent. This king(i.e.Udayana), thinking of him with affection, desired to have an image of his person, therefore, he asked Maudgalyānaputra, by his spiritual power, to transport an artist to the heavenly mansion to observe the excellent marks of Buddha's body, and carve a sandal-wood statue. When Tathāgata returned back, the carved image of sandal-wood rose and saluted the Lord of the world".[41]

Huien-Tsang has given one more account regarding the image of the Buddha which is as follows:

"Formerly, when Tathāgata ascended into the Tushita heaven to preach the Dhamma for the benefit of his mother, Prasenjit-rājā, having heard that the king Udayana has caused a sandal-wood figure of Buddha to be made, also caused this image to be made".[42]

---

40. H.A. Giles, *The Travels of Fa-Hien* (Tr.), Cambridge, 1923 (reprint), London, 1966, pp. 30-31.

41. S.Beal,*Buddhist Records of the Western World*, Vol.I, London, 1906,(reprint in Delhi,1969), pp.235-36,

42. S.Beal. *Hiuen-Tsang*, Delhi,1969,p.4ff.

The first account of Huien-Tsang regarding the making of a Buddha image caused by similar circumstances but with a difference. According to Fa-Hien's account it was to be commissioned by king Prasenjit whereas according to Huien-Tsang by king Udayana of Kausambi. Despite the Buddha image being of sandal-wood, it mentions of a stone canopy. According to Przyluski's study on *Aśokāvadāna*, Kausambi came into prominence after the 1st Buddhist council and this emphasis on Kausambi in the Pāli canon probably took place after Aśoka's life time.[43] It is also said that king was least interested in Buddhism.[44] Another account provided by Huien-Tsang seems to be a variation of Fa-Hien's account that Prasenjit makes a sandal-wood image of the Buddha after having heard of Udayana's image. It is true that Huien-Tsang visited Śrāvastī and also the monastery was visited by Fa-Hien which was in ruins where he found an image.[45] Cunningham who had excavated the site , identified this as temple number-3 which was seen by Huien-Tsang and which contained a 'Bala' type image.[46] So, there is a clear difference between the two accounts. Fa-Hien mentions of a seated image whereas Huien-Tsang describes standing image.

Apart from these accounts regarding the early images of the Buddha, there is a literary source, found in the *Vaṭṭangulirāja Jātaka*, 37th *Jātaka* of the *Paññāsa Jātaka*, which relates the almost similar story regarding the Buddha image as mentioned above. Padmanabh S.Jaini,[47] in his article, 'On the Buddha image', has referred the *Vaṭṭangulirāja Jātaka* (Burmese) which mentions that the image of the Buddha was made during His life time by Prasenjit. But about this *Jātaka* it is said that it was probably originated in the 13th or 14th century and was popular only in Burma, Thailand and Cambodia and has no counterpart in India and Ceylon.[48] Here it is notable that all these three sources are provided from the foreign accounts and are

---

43. Jean Przyluski,*Legends of the Emperor Aśoka in Indian and Chinese Texts*, tr. by D.K.Biswas, Calcutta, pp.78-80.

44. Niti Adval, *The Story of king Udayana*, Varanasi,1970, pp.203- 223.

45. S.Beal, *Buddhist Records of the Western world...* p.235.

46. *Annual Report, ASI*, Vol.XI, Calcutta,1880,pp.86-88.

47. Padmanabh S.Jaini,*op.cit.*, pp.184-186.

48. *Ibid.*,p.184.

comparatively very late. These evidences support the practice of image worship and its sanction comes from the Buddha himself. It seems that this type of idea might have evolved by the Theravādins as a justification for image worship.

There are some other sources regarding the early images of the Buddha. According to *Ekottarāgama* (Mahāsaṅghika counterpart to Pāli *Aṅguttara Nikāya*) "Udayana makes an image of sandal-wood five feet high and Prasenjit follows suit with a golden image".[49] Another account in Tibetan repeats much as the Huien-Tsang's version (Udayana image) except that the image is depicted in a standing position and that a dazzling 'sun stone' was placed in front of the Uṣṇīṣa.[50] A later Tibetan commentarial literature mentions that the first image of the Buddha was made on the request of Anāthpiṇḍaka to add splendour to the gathering of monks at a noon-meal held in the absence of the Buddha.[51]

1.4. Tibetan and related sources regarding the early images of the Buddha:

Tibetan sources can be divided into two parts. Part one deals with the images of the Buddha that were made just after His demise and part two deals with the images which were made during the Aśokan period onwards.

Part one:

Tibetan historian Tāranāth[52] claims that the image of the Buddha was made just after His demise. In his account he mentions that there were three Brāhmaṇa brothers, Jaya, Sujaya and Kalyāna being converted to Buddhism, they built temples at Vārānasi, Rājagṛha and Bodhgayā respectively. Once the youngest son Kalyāna shut himself,

---

49. N.Dutta, *The spread of Buddhism and Buddhist schools*, Delhi, 1980, p.130; A.Waley.Did the Buddha Die of Eating Pork"' *Melanges Chinois et boudhiques*, Vol.I,1932,p.352-4.

50. Loden Sherap Dagyab,*Tibetan Religious Art* (Asiatische Forschungen, Vol.52), part I, Wiesbaden,1977,p.22.

51. *Ibid.*

52. Tāranāth, *History of Buddhism in India*, Tr. by Lama Chimpa and Alaka Chattopadhyaya(Ed.) Simla, 1970,pp.31-42.

along with the artist and materials, for seven days in the Mahābodhi temple at Bodhgayā, in order to carve a Buddha image, but on the sixth day was forced to open the temple on account of his mother's threat to kill herself if she was not shown the Tathāgata image. She forced the door open on the sixth day to determine the likeness of the Buddha image to the Tathāgata whom she had seen in her lifetime. A similar account of the image made by youngest of the three brothers occurs in the Biography of Dharmasvāmin,[53] as Tibetan pilgrim who visited India in 13th century A.D. These Tibetan sources seem to be myth because anyone going through the accounts of Tāranāth and Dharmasvāmin in detail would be able to judge the authenticity of the narratives. First of all they were in the 13th century A.D. and describing about years' back without mentioning any concrete evidence and it is quite clear that the Mahābodhi Temple at Bodhgayā and the image in it were made during the Gupta priod as is also evidenced from the accounts of Fa-Hien and Hiuen-Tsang.[54] On the basis of archaeological excavations which were first of all made by Cunningham,[55] the earliest construction pertains to the Aśokan period and not before that. So, the accounts of both Tāranāth and Dharmasvāmin are merely legends.

Part Two:

Tāranāth further mentions about the two images of the Buddha which Aśoka saw during the process of his conversion[56] and other seen by him (Aśoka), during the period of his conquest on Kaliṅga, before his conversion.[57] Fa-Hien in his account mentions that the Buddha image was made at Sāṅkāsya by Aśoka along with a shrine and a pillar behind the shrine.[58] He further mentions that he saw only a shrine, neither the Aśokan shrine nor the image enshrined by

---

53. George Roerich and A.S.Altekar, *Biography of Dharmasvamin*(Tr.),Patna,1959, pp.67-69.
54. T.Watters,*On Yuan Chwang's Travel in India*, Vol.II, London.1906,pp.113-36.
55. D.Mitra, *Buddhist Monuments*, Calcutta,1971,p.61.
56. Tāranāth, *Op.Cit.*,p.39.
57. *Ibid.*,p.75
58. Fa-Hien...*Op.Cit.*,p.25.

him.[59] But none of the above references have been supported by archaeological evidences.

1.5. Factors responsible for the absence of the Buddha images in early Buddhist art.

The representation of the Buddha in aniconic form instead of anthropomorphic, in early Buddhist art at Bharhut, Sānchi, Bodhgayā and so on, before the Kuṣāṇa period, had initially intrigued scholars. The Buddhist community was well aware of the description of the Buddha's personality from Pāli literature and the thirty-two auspicious physical marks. His image did not appear for nearly four hundred years in early Buddhist art.[60] The aniconiic nature of early Buddhist art was initially taken as incapacity of Indian sculptors, both in terms of idea as well as technique. This view was refuted by Coomaraswamy[61] and other scholars.[62] Actually it was not the incapacity of Indian sculptors but this particular mode of art - expression was deliberately bounded by the norms of early Indian religious art, especially Buddhist. Various causes for the absence of the Buddha image in early Buddhist art have been suggested by scholars.

Aniconic mode of representation in early Buddhist art was keeping with the contemporary attitude in higher religions and its tradition. The sixth century B.C. was an epoch of great religious ferment of philosophical enquiry. The whole Indian religious scene was divided into two main currents; Brāhmaṇa and Śramaṇa. The Vedic Brāhmaṇa religion was essentially henotheistic in which sacrifice (*Yajña*) through fire played a dominant part. They conceived their deities aniconically.[63] Hence, they had no need for image making and worship. Buddhism originated after later Vedic period and the second

---

59. *Ibid.*, p.26.
60. Albert Grunwedel, *Buddist Art in India*, London,1901,pp.161-62, S.K.Gupta,"Causes of the absence of Buddha image in early Indian Art",*K.P.Jaiswal Commemoration* Volume,Patna,1981,p.134.
61. A.K.Coomaraswamy,*The Origin of the Buddha Image..* pp,33-38.
62. Victor Goloubew, Quoted in D.L. Snellgrove (Ed.), *The Image of the Buddha...,*p.59.
63. K.K. Dasgupta, *Comprehensive History of India*, (Ed.), vol.III, part II, Delhi, 1982, p.857.

current emerging within the Vedic fold at that time was that of the *Upaniṣads*.[64] They criticised the old Vedic sacrificial religion, efficacy of its over-grown ritualism and its various propositions concerning cosmos.[65] The Upaniṣadic seers were engaged in metaphysical and philosophical enquiries about *Brahma* through supernatural insight gained from penance and meditation. They also had no need for images. Hence, the Vedic religion, one of the higher religions at that time in India with mass base, was opposed to iconism. The Śramaṇa front, to which Buddhism and Jainism belonged, was as great as the Vedic Brāhmaṇa tradition, as patronage and intellectual leadership came from the higher strata of society. It is notable here that over 40% of the elite monks and nuns listed in *Theragāthā* and *Therigāthā* belonged to the Brāhmaṇa caste, 29% to the Vaishya and 22% to the Kshatriya varṇas. So, due to their previous background too they followed the same mode of aniconism.[66] On the whole image making and worship had not in general become a part of the milieue in most part of India, at least not in the higher classes of society so far as the deities of higher tradition were concerned.[67] The early representation of deities were by means of symbols which atleast in orthodox or circle of higher religion was universal and that is why Coomaraswamy has rightly called it an "iconography without icons".[68]

References regarding the image making in pre-christian era have been gleaned from the texts of Pānini, Patañjali, Megasthenese, Gautama's *Dharmasūtra*, *Grihya Sūtra* and archaeological as well as numismatic evidences also testify to it.[69] The list of deities whose images were made hardly included deities from the 'higher' orthodox Vedic pantheon[70] or such deities who till then had not been accepted

---

64. A.L. Basham, "The Background to the rise of Buddhism" in A.K. Narain (Ed.), *Studies in History of Buddhism.* Delhi, 1980, pp. 13-14.

65. *Ibid*, p. 14.

66. B.G. Gokhale,"The Early Buddhist Elite," *Journal of Indian History*, 1965, Vol. XLII, part II, pp. 391-402.

67. J.N. Benerjea, *The Development of Hindu Iconography,*, Calcutta, 1956, Chapter-II and III, especially pp. 75-77.

68. A.K. Coomaraswamy, *The Origin of the Buddha Image...*, pp. 4-11.

69. K.K. Dasgupta, *Op. Cit.*, pp. 857- 859.

70. A.K. Narain, "First Image of the Buddha and Bodhisattva; Ideology and Chronology...",(Ed), *Studies in Buddhist Art of South Asia*, Delhi, 1985, p.2.

by higher religions. Patañjali who flourished in 2nd century B.C. has called Śiva, Skanda and Visākha as *Laukika* devatās or folk deities.[71] The information provided by above sources suggest that deities whose images were made mostly related to deified local folk and guardian deities and divinities of exotic origins, and deities not from the higher orthodox religions. The above facts tend to prove that in higher religions image-making was not in vogue except in local cults which were popular among the masses.

Apart from the nature of contemporary religions and religious art tradition, a clear cut ideological obligation arising out of the Buddha's words and the creed of orthodox Theravāda were the deterring factor in image making of the Buddha. Coomaraswamy has rightly remarked that the artists of pre-Kuṣāna period were not incapable to carve the anthropomorphic images of the Buddha. According to him craftsmen who were capable of producing Parkham Yakṣa and Patna Yakṣi images and the reliefs at Bharhut and Sānchi would have had no difficulty in representing Gautama in human form, if they had been required to do so.[72] So more than the art tradition it was the Buddhist interdiction which was responsible for aniconic mode of representation in early Buddhist art, though, the interdiction regarding the Buddha image have nowhere been explicitly mentioned in early Pāli canonical literature. However, there is one indirect reference to the prohibition of the making of the Buddha image. In a chapter of Sarvāstivādin *Vinaya* which deals with "the decoration of monasteries, the Buddha is asked by Anāthapiṇḍaka, "world-honoured one, if images of yours are not allowed to be made, pray, may we not at least make images of Bodhisattvas in attendance upon you"? The Buddha then grants permission to the request."[73] Apart from this there is one more reference regarding the canonical interdiction for the making of the Buddha image, which have been mentioned in the literary section of this chapter, but that one is of later origin. Apart from this, the various metaphysical and religious

---

71. K.K. Dasgupta, *Op. Cit.* p.857.

72. A.K. Coomaraswamy, *op. cit.,* p.15.

73. A. Waley, appended to "Did the Buddha Die of Eating Pork", *Melanges Chinois et boudhiques*, Vol. I, 1932. pp.352-4, quoted from A.K. Narain (Ed.), *Op. Cit.,* p. 27.

ideas concerning the Buddha-nature and his words, scattered in Pāli literature, have been put forth by scholars as the possible explanation for the absence of the Buddha image. Overall they reflect the Buddha's idea and early Buddhist attitude towards iconism.

In the Theravāda tradition Buddha Sākyamuni was conceived as a man who due to his own efforts became the Buddha. The Theravādins did not regard the Buddha as *Lokottara* but attributed to him almost all powers and qualities of a *Lokottara*.[74] The Buddha always equated and identified himself with the Dhamma.[75] When the monk Vakkali on his death bed earnestly desired to see the Buddha in person, the Blessed one said, "*Alam Vakkali kim te pūtikāyena diṭṭhena. Yo kho Vakkali dhammam passati so mam passati, Yo mam passati so dhammam passati*", i.e., He who sees Dhamma sees me , he who sees me sees Dhamma.[76] The identification of Dhamma with the Buddha in his life time and the sole guidence of Dhamma to the Buddhist community after His *Mahāparinirvāṇa* placed the Dhamma in an advantageous position than the Buddha since the real body of the Buddha was supposed to be the *Dhammakāya* (collection of Dhamma) which is materially abstract, hence he could not be represented in anthropomorphic form.

The above mentioned contemporary religious art tradition, and Buddhist interdictions gleaned from early Pāli canonical literature has been proposed by scholars as the possible cause of the absence of Buddha image in early Buddhist art. It has been pointed out that even after one or two centuries of the Buddha's demise the Theravādins retained the human conception of the Buddha as a saint and teacher per-excellence.[77] So, even after two centuries of His demise, the Buddha image (in human form) could not have come into being depicted in the monuments of Bharhut, Sānchi and Bodhgayā, which only manifest the symbolical representations of the Buddha.

---

74.   N.Dutta *Buddhist Sects in India*, Delhi, 1978, pp. 98-103.
75.   N.Dutta, *Mahāyāna Buddhism*, Delhi, 1973, pp. 144-45.
76.   *Ibid.*
77.   S.K. Gupta, *Op. Cit.*, p. 135.

1.6 Causes of the emergence of the Buddha image:

There was no dearth of devotion towards the Buddha in early Buddhism which manifested through various symbols viz., *Bodhi*-tree, wheel, *Stūpa* and other uniconic representations as appeard on the external decoration of the *Stūpa*. The anthropomorphic images of the Buddha came into existence between first century B.C. and A.D. Now the question arises : when the *Stūpa* was already providing a suitable cult object, what was the possible reason that necessitated the making of the Buddha image?

To trace the possible causes of the emergence of the Buddha image, we have divided our subject of discussion into two sections: A and B.

A. A study of the ideological development in Buddhism in the purview of the emergence of the Buddha image:

Part-I: Conception of the Buddha in different schools of Buddhism:

Historically the Buddha was born in 563 B.C. at Lumbini grove and named as Siddhārtha Gautama. From his early childhood he showed a meditative bent of mind. At the age of 35 he attained knowledge at Bodhgayā and from this time onwards he began to be called the Buddha or the enlightened one. [78] In course of time when Buddhism was divided into different schools, the Buddha was conceived by different schools in different ways.

THERAVĀDA: The Theravāda or Sthaviravāda which is the oldest and orthodox school of Buddhism, held the view that the Buddha was a human being but he possessed certain super human qualities. In the *Dīghanikāya* the Buddha is described as an *Arahat*, a fully awakened one, endowed with knowledge and good conduct, a happy knower of the world and a teacher of men and his gods.[79] However, even being endowed with supernatural powers his physical body was subject to physical laws as any mortal and in the *Nikāyas*

---

78. N.Dutta, *Buddhist Sects in India*, Calcutta,1970,p.231.
79. *DN*,III i, 135.

the Buddha has called his body *Putikāya* i.e. a body full of impure matters.[80]

In the Theravāda the Buddha was also addressed as Bhagawān and was worshipped due to his supernatural powers, achievements and unsurpassed knowledge. Despite stress on Dhamma, the ethical teaching and salvation through one's own efforts, devotion towards the Buddha similar to gods was not lacking in early Buddhism.[81] But being aware of the people's devotion towards him, the Buddha always discouraged the monastic community as in the case of Vakkali. But so far as the laity was concerned the Buddha gave them his hair, nails, robes, and other corporeal relics as mementos and the laity used to make *Stūpas* over them so that they could worship it to satisfy their devotional cravings.[82] Even in his last days when asked upon by Ānanda as to how he should be honoured after his demise, the Buddha instructed Ānanda to make *Stūpas* on the crossing of four roads over his corporeal relics and worship it. The Buddha also gave instruction to undertake pilgrimage to the places related to the different events of His life,e.g. Lumbini (place of birth), Bodhgayā (place of enlightenment), Sārnāth (place of first preaching), Kuśinagara (place of Mahāparinirvāṇa).[83] In this way, we find that he did not discourage devotion towards Him and being a practical compaigner gave a rational outlet to such feelings by placing symbols and objects as a substitute for the act of devotion towards Him.

MAHĀSAṄGHIKA: The Mahāsaṅghika school came into existence just after the 2nd Buddhist council held at Vaishali. The school is also known as the precursor of Mahāyāna Buddhism. The main deviation made by this sect was that they deify the Buddha and assert that the Buddhas are *Lokottara* (supramundane) and are connected only externally with the worldly life.[84]

The Mahāsaṅghikas laid special emphasis on the Buddha-*Bhakti* and enunciated that devotion and worship of the Buddha and of the

80. *Cullavagga*, p.293;G.S. Mishra, *The Age of Vinaya*, p. 60, fn.132.
81. B.M.Barua,"Faith in Buddhism" in B.C.Law(Ed.), *Buddhist Studies*,p.32.
82. D.Mitra, *Buddhist Monuments*,Calcutta,1980,p.23.
83. H.Kern, *Manual of Indian Buddhism*, Delhi,1972,p.43.
84. R.C.Majumdar,(Ed.), *The Age of Imperial Unity*, Bombay,1953,p.381.

Stūpas which contain the Buddha's relics is the means not only to accomplish mundane happiness but also to attain *Nirvāṇa*. Here the Buddha is endowed with unlimited divine powers He untiringly endeavours to enlighten the sentient being and He has unlimited compassion. [85] So, with the rise of the Mahāsaṅghikas the Buddha was gradually regarded as god and the tradition of stressing on the Buddha-*Bhakti* to achieve one's goal came into existence.

SARVĀSTIVĀDA: The Sarvāstivāda branched off from the Theravāda before or after the 3rd Buddist council held at Pāṭaliputra. The Sarvāstivādins enjoyed their position in Mathura, Gandhāra and Kashmir regions. The Sarvāstivādins held the view that the Buddha was a historical person. He was not beyond infallibility, and had nothing extraordinary about Him. Like Theravādins, they conceived the Buddha as a human being, who, by prolonged meditation attained Buddhahood and duly became omniscient at Bodhgayā. [86] Unlike the Theravādins the Sarvāstivādins attributes to the Buddha divine, sometime superdivine powers.[87]

The Avadāna literature which is supposed to be the works of Sarvāstivādins[88] reflect a full-fledged concept of devotionalism. The Avadāna literature made a strong appeal to the people by its stories relating to the Buddha, his predecessors, disciples, *Arahats*, devotees etc. by their glorification of the Buddha and Bodhisattvas and eulogising the acts of his disciples and devotees. The *Avadāna Sataka* and the *Divyāvadāna* are the most prominent *Avadāna*s which by their narratives and legends inculcated devotion to the Buddha and to other holy personages. The *Divyāvadāna* begins with the Mahāyānistic benediction,"Om, adoration to all the Buddhas and Bodhisattvas", though it belongs as a whole to Hīnayāna.[89]

---

85. N.Dutta, *Op.Cit.*, pp. 75-80;*Mahāvastu Avadāna*,II,362 *ff*.

86. A.N.Lahiri,"The Sarvāstivāda" : Its inherent vitality and widespread popularity."(Paper read in a Seminar in Delhi University,1986, pp. 3-4)

87. N.Dutta.*Op.Cit.*, pp. 169-170.

88. H.Nakamura, *Indian Buddhism*, - Japan,1980,p.84.

89. M.Winternitz, *A History of Indian Literature*, Vol.II,Delhi,1972(2nd ed.), p. 284.

In this way, we find that the Sarvāstivādins on the one hand conceived the Buddha as a human being and on the other hand, they glorify the Buddha and laid stress on devotion to Him.

Part-II: The Concept of Kāya:

Generally the word *Kāya* stands for body. But in Budddhist terminology *Kāya* means the two or three—fold bodies of the Buddha. The evolution of Kāya theory gave impetus to the deification of the Buddha. The description of the *Kāya* concept can be traced even in early Buddhist literature. Actually different schools of Budddhist thought conceived the Buddha in different ways and perhaps due to the convenience of the interpretation of their own view they divided the conception of the Buddha's body into two or three parts viz.,(1) *Dhammakāya*. (2) *Rūpa or Nirmāṇakāya*, and (3) *Sambhogakāya*.

Let us survey the concept of *Kāya* in different schools of Buddhism:

THERAVĀDA : The Theravādins evolved the concept of two-fold bodies of the Buddha viz., *Rūpa* or *Nirmāṇakaya*, and *Dhammakāya*. According to Theravādins the real body of the Buddha is *Dhammakāya*(collection of His Dhammas), which cannot be seen because it has no form. The *Dhammakāya* is eternal without appearence or disappearance. [90] It seems that the concept of *Dhammakāya* might have evolved to emphasize the teachings of the Buddha.

Rūpa or Nirmāṇakāya is the created body in which the Buddha rendered his services for the welfare of human beings.[91] But the Buddha called this body (*Rūpa* or *Nirmāṇakāya*) as *Putikāya* i.e. a body full of impure matters.[92]

It seems that since the created body of the Buddha is subject to physical laws as any mortal and the *Dhammakāya* i.e. collection of Dhammas which is obviously eternal. This is why *Rūpa or Nirmāṇkāya* is called *Putikāya* and the *Dhammakāya*, real body of the Buddha.

---

90. N.Dutta, *Aspects of Mahāyāna Buddhism and its relation to Hīnayāna*, London,1930,pp.97-102.

91. *Ibid.*

92. *Cullavagga*,p.293.

SARVĀSTIVĀDA : The Sarvāstivādins who retained the realistic conception of the Buddha differed a little from the Theravādins. The *Kāya* conception of this school is depicted in the *Divyāvadāna,Lalitavistara,* and *Abhidharmakośa.*

The concept of *Rūpa or Nirmānakāya and Dhammakāya* as depicted in the *Divyāvadāna* are similar to those of the Theravādins.

The *Lalitavistara* which is considered to be the complete biography of the Buddha, depicts Him as superhuman being. In this text the Buddha has been deified but there is no trace of the *Trikāya* conception in it. In the original form, the *Lalitavistara* depicts the Buddha as a human being with superhuman attributes. But in the later part of the book, where the Buddha's attributes are mentioned, He is called as great tree (*Mahādruma*), because He possesses a body of *Dharmakāyajñāna* (the knowledge of *Dharmakāya*); this concept is very likely a Mahāyāna addition. [93]

The *Abhidharmakośa* of Vasubandhu replaces the concrete conceptions of the *Dharmakāya* found in the *Nikāyas* and the *Divyāvadāna* by an abstract one. According to Vasubandhu, *Dharmakāya* means the quality adhering to the Buddha as the purified personality.[94]

The concept of *Rūpa or Nirmānakāya* as depicted in the text is almost similar to those of the Theravādins. According to Vasubandhu the physical body (*Rūpa or Nirmānkāya*) of the Buddha does not show much outward change after attaining the enlightenment. The *Rūpakāya* of the Buddha is in fact the *Rūpakāya* of Bodhisattva and hence it is *Sāsrava* (impure).

MAHĀYĀNA: It was the Mahāyānists who incorporated the idea of *Trikāya* theory in Buddhism, though they inherited the concept of *Nirmānakāya* from the Mahāsanghikas.

According to the Mahāyānic ideas the Bodhisattvas after acquiring all the necessary Dharmas and practicing the *Prajñāpāramitā* become *Sambuddha*. He then rendered services to the beings of all *Lokadhātus*(worlds) of the ten corners at all times. The body through which the Buddha rendered services is known as *Nirmānakāya*. Here,

---

93. N.Dutta, *Op.Cit.,* pp. 103-104
94. *Ibid,* pp.104-108.

we find the Mahāyānic concept of *Rūpakāya* as different from the Theravādins who called it *Putikāya.*

The Mahāyānists define the concept of *Sambhogakāya* in this way : Bodhisattvas, after attaining the *Bodhi* by means of the *Prajñāpāramitā*, take a body endowed with thirty-two major and eighty minor signs, with a view to preaching the doctrines of Mahāyāna to the Bodhisattvas and at the same time to arouse in their minds joy, delight and love for the excellent Dharma. This type of the Buddha's body is known as *Sambhogakāya.* The *Sambhogakāya* is so called because it is shown as the special acquisition of the Buddha on account of merits accumulated in several lives[95] and he enjoys the bliss of knowledge through this body. However, some of the Mahāyānic works do not distinguish the *Rūpa or Nirmāṇakāya* from the *Sambhogakāya.*[96]

About the *Dhammakāya*, the Mahāyānists say that it is devoid of all marks and is inexpressible. It possesses the eternal, real and unlimited *guṇas* and it has neither *Citta* nor *Rūpa*, and it is not different from them. It is said that the Buddhas may have their individual *Sambhogakāyas* but all of them have one *Dharmakāya.* [97]

Part-III : Evolution of Bhakti-cult in Buddhism :

The root of Bhakti in Buddhism can be traced from the early period through *Triśaraṇa* formula in which the novices take refuge in the Buddha, the Dhamma, and the Saṅgha. It is quite obvious that without showing devotion or respect towards the Buddha, one can not take refuge in Him. However, there has been a good deal of discussion among scholars about the meaning of the term *Śaraṇa.* [98] The word *Śaraṇa* is used in the general sense of protection. A roof gives *Śaraṇa* against rain, and a parasol against the Sun. In the religious sense the protection is not from the worldly evils but it helps a man to defend himself against the evil impulses which produce suffering in this world and the next and such protection can be

---

95.   *Ibid.*, pp.123-124.
96.   *Ibid.*, pp. 119-120.
97.   *Ibid.*, pp. 123-124.
98.   N.Dutta, *Early Monastic Buddhism*, vol. II,Calcutta, 1941, p. 277.

secured by developing faith in the Buddha, the Dhamma, and the Saṅgha.[99]

Gradually the Buddha became an object of adoration and meditation in Buddhism as shown in the formula of "Adoration to Buddha" (*namo tassa Bhagavato Arahato sammā sambuddhassa*). The term "*Buddhānusmriti*" in early Buddhist literature had four meanings: 1. meditation on the virtues of the Buddha; 2. hearing the name of the Buddha; 3. repetition of the name of the Buddha; and 4. meditation on the figure of the Buddha.[100] It seems that the Buddha had not completely rejected devotion shown to him. The *Majjhimanikāya* mentions that the Buddha himself had said," even those who have not entered the path are sure of heaven if they have love and faith towards me.[101]" The word Bhagavān occured several times in early Buddhist literature and in Aśokan inscriptions as in the case of Piparahwa, Lumbini[102] etc. Ironically, we never come across the term '*Bhakti*' in early Buddhist canonical literature, only the act of devotion is expressed. The *Aṣṭasāhasrikāprajñāpāramitā Sūtra*, one of the earliest Mahāyānic texts, reflects devotionalism as a medium to gain merit and this can be done by worshipping the Buddha image as the text explicitly mentions that the purpose of making an image was to cause men to gain merit by seeing the Buddha (*Buddhadarśanapuṇya*) and nothing else.[103]

The *Saddharmapuṇḍarika Sūtra*,[104] another early Mahāyānic text also speaks of Buddha-Bhakti, cult of relics and image worship. The text mentions that those who offer *stūpas*, those who make or cause to be made images, carvings, coloured images (paintings), children even in plays who draw the Buddha images, or those who make offerings to those images of various sorts, any of these will achieve

---

99. A.L.Basham,"The Background to the Rise of Buddhism" in A.K.Narain(Ed.), *Studies in History of Buddhism*, Delhi,1980,p.20.

100. H.Nakamura, *Op.Cil.*, *p. 84*.

101. *M N.*, I. 142 (*Yesam mayi Saddhamattam pemamattam saggaparayana ti*).

102. D.C.Sircar,*Select Inscriptions*, Vol.I. Calcutta, 1942, pp. 84,70.

103. A.K.Narain (Ed.) *Studies in Buddhist Art of South Asia*(the relevant portion of the text is discussed in the book), Delhi, 1985, pp. 46-48.

104. M.Winternitz., *A History of Indian Literature*, Delhi, 1972,pp.297-304.

the Buddha-path through the use of those expedient devices of the Buddha.

In this way, we find that *Bhakti*, earlier whose purpose was only to express reverence to the teachers, elders etc., now became a cult.

B. A Survey of the factors contributing to the emergence of the Buddha image.

(1) It is well known that early Buddhism was basically an ethical and moral religion. The Buddha said that man himself is the master of his own destiny. The door of the Buddha was open to all. Each man must tread the road and win the goal for himself, none helping or hindering him in the supreme task. So goal can be achieved through one's own efforts. There was no room for adoration or worship to gain the merit.

However, gradually a tendency evolved in Buddhism which started to regard Gautama Buddha as more than a human being but not as a divine being. Most probably under the impact of this changing tendency, symbolical worship was started in Buddhism.

With the appearence of the Mahāsaṅghikas, a revolutionary change took place in Buddhism. They made the Buddha *Lokottara* (supramundane) and laid special emphasis on the Buddha-*Bhakti*. The Mahāsaṅghikas also enunciated that through devotion and worshipping the Buddha one can also attain the highest goal.

The Sarvāstivādins, though, conceived the Buddha as a historical person, attribute to Him divine and superdivine powers. They introduced the Avadāna literature to glorify the Buddhas and Bodhisattvas. The Avadāna also laid stress on devotion to the Buddha and other holy personages.

The concept of *Trikāya* is said to have evolved to introduce image worship in Buddhism. As discussed earlier, according to the Theravadins the real body of the Buddha is *Dhammakāya* (collection of Dhamma) and *Rūpa or Nirmāṇankāya* (i.e. the physical body of the Buddha through which He rendered His services) is *Putikāya* i.e. a body full of impure matters. So, the Buddha's physical body should not be worshipped. But, when devotionalism came into existence in Buddhism and obviously the Buddha was not present in person at that time then the devotees might have felt the need for the concrete representation of the Buddha to whom they can express their devotion and then the process of the deification of the Buddha might have started. Thus, it reveals that the Theravādins, the Mahāsaṅghikas and the Sarvāstivadins consider the *Dhammakāya*(collection of Dhammas) as the real body of the Buddha. But like the Theravadins, the Mahāsaṅghikas do not consider the *Rūpa* or *Nirmaṇākāya* of the Buddha as the *Putikāya*, however, the latter ones believe that the *Rūpa or Nirmāṇakāya* is the self created body of the Buddha by which the Master rendered his services to the cause of humanity. Later on, the Mahāsaṅghikas, the Sarvāstivādins etc. glorify the body of the Buddha which contributed to the deification of the Buddha and which finally led to the emergence of the Buddha images.

Thus, we may say that with the evolution of *Bhakti*-cult and the concept of *Trikāya* theory in Buddhism, the Buddha became deified which led to the emergence of the Buddha image.

(2) We have discussed above that the evolution of *Bhakti*-cult in Buddhism was the prime factor for the emergence of the Buddha image. It is said that it was the Brāhmanic *Bhakti* movement which permeated the idea of devotionalism in Buddhism. The Brāhmanic *Bhakti* movement was at its peak during the Kuṣāṇa period[105] when the images of the Buddha were made for the first time.

Service, devotion, loyalty and attachment to the deity were the characteritics of the *Bhakti* movement resulting in the making of images and building of temples for adorable beings and deities. It was in this spirit that Heleodoros installed a Garuḍa capital as a token of his respect for Vāsudeva at Vidisa in early 2nd century B.C. and calls himself as Bhāgavata and addresses his deity as Bhagavān Vāsudeva. Under the impact of this movement the worship of Śiva, Skanda, Visākha, Śaṅkarṣaṇa and Vāsudeva came into vogue about the

---

105. R.G.Bhandarkar, *Vaiṣnavism, śaivism and other minor sects*, pp. 9-10.

2nd century B.C.[106] The images of these divinities have been discovered in Mathura belonging to this period. The images are both in stone and terracotta.[107]

There is great affinity between the *Saddharmapuṇḍarika Sūtra* and the Brāhmanic text *Gītā* in terms of devotion or *Bhakti* and on the basis of this we may say that Buddhism inherited the idea of *Bhakt* from Brāhmanism. Let us compare both the works about the similarity of terms.

In the *Gītā* Kṛṣna says :

" I am the support, father, mother and grandfather of the world"...."I am the goal, sustainer, the Lord, the witness, the abode, the refuge and the friend". (*Pitāḥmasya jagato māt pitāmaḥ....gatirbhartā Prabhu sākṣī nivāsaḥ śaraṇaṁ suhrit*)

(*Bhagavad Gītā*, IX.17-18)

In the *Saddharmapuṇḍarīka Sūtra* the Buddha proclaims:

"I am the father of the world, the self-born, lord of all beings and curer of ill". (*Yameva 'haṁ loka-pitā svayaṁbhūḥ cikitsakaḥ sarva-prajāna nāthaḥ*) (*Saddharmapuṇḍarīka Sūtra*, XV.21)

Again Kṛṣna says in the *Gītā* :

" I am alike to all beings, none is favoured or disfavoured by me". (*Samo'haṁ sarbhutesu na me dveṣyo'sti na priyaḥ*) (*Bhagavad Gītā*, IX.29).

A similar thing the Buddha says in the *Saddharmapuṇḍarika*:

" I do not have to please anyone. I do not favour or disfavour any". (*Anunīyatā mahya na kā-cid asti premā ca doṣas ca na me kahiṁ cit.*) (*Saddharmapuṇḍarīka*, V.22)

The date of the *Saddharmapuṇḍarīka Sutra* has been fairly settled. It was translated into Chinese by Dharmarakṣa in 286 A.D. which indicates that by the third century A.D. it had established its eminence.

---

106. D.L.Snellgrove(Ed.), *The Image of the Buddha*, Delhi, 1978,p. 48.

107. Devangana Desai,"Social Background of Ancient Indian Terracottas,"in D.P.Chattopadhya (Ed.),*History and Society*, Calcutta, 1978,p. 159;Sharma.*Op,Cit.*, p.135.

Most of the scholars like, N.Dutta,[108] Winternitz[109] P.L.Vaidya[110] Narendra Deva,[111] etc. opined that the work was first composed in the first century A.D. which indicates the approximate time when the element of *Bhakti* combined with selfless services have taken a full fledged position in the Buddhist tradition.

The date of the *Bhagavadgītā* is not beyond controversy, however, it is believed that it was composed sometime between the 5th and 4th century B.C.[112] So it proves that the element of *Bhakti*, with emphasis on disinterested action was first developed in the Bhagavata school of Brāhmanism or Hinduism, which later on was adopted by the Mahāyāna Buddhism in its own setting as confirmed by the comparision of the *Saddharmapuṇḍarika* with the *Bhagavadgītā*. However, only on the basis of similarity between *Bhagavadgītā* and *Saddharmapuṇḍarika* in terms of devotinalism, it can not be accepted that Brāhmanic Bhakti movement was solely responsible for infusing devotionalism into Buddhism. Because, as discussed above, the element of *Bhakti* was already prevalent in Buddhism from the days of the Buddha and it was under its impact that symbolic worship was started in Buddhism. Of course, the contribution of Brāhmanic Bhakti movement was that it gave impetus to infuse the personified form of devotion in Buddhism and which led to the emergence of the Buddha image.

(3) The popularity of the cult of Yakṣa, Yakṣi, and Nāga, Nāgin among the general people is also credited for their contribution to the emergence   of the Buddha image.[113] The early images of the Buddha/Bodhisattva discovered from Mathura show a great affinity to the images of Yakṣa.[114] Some of the images of Buddha/Bodhisattva also show affinity to the Nāga images.[115]

---

108. N.Dutta,*Saddharmapuṇḍarika(Ed.)*, Calcutta, 1953,Preface,p.17.

109. M.Winternitz, *A History of Indian literature*, Vol. II, Delhi,1963, pp.303-4

110. P.L.Vaidya, *Saddharmapuṇḍarīka* (Ed.) Darbhanga, 1960, Preface, p. 14.

111. Narendra Deva,*Baudhadharmadarśana*, Patna, 1971, p. 142.

112. K.N.Upadhaya, *Early Buddhism and Bhagavadgita*, Delhi,1971, pp. 16-29.

113. D.L.Snellgrove(Ed.), *The Image of the Buddha*, Delhi, 1978, pp.47-48.

114. A.K. Commaraswamy, *Origin of the Buddha Image*, Delhi, 1972 (reprint), pp. 2,3,4,48,50.

115. *Ibid*. pls.48,49,50.

The tradition of worshipping the images of Yakṣa, Yakṣi and Nāga, Nāgin is quite old. The term Yakṣa appeared quite often in early Indian literature. The Yakṣas are associated with both benevolent and malevolent aspects and believed to have possessed of magical or supernatural powers; their female counterparts, Yakṣi is more famous for withcraft. In the *Atharaveda* the subjects of Kubera, i.e. Yakṣas and Rākṣasas, are called as *Punyajanas*.[116] The *Gītā* mentions that the people of Rājasika class used to worship Yakṣas and Rākṣasas.[117] It is also held that the Rākṣasas had two-fold divisions, the one as guardians of treasures was known as the Yakṣas while the other class, which was notorious by nature and disturbed the sacrifices, was called Rākṣasas.[118] Generally the Yakṣas are supposed to be friendly with human being while the Rākṣasas are of evil nature. Gradually Yakṣas gained much popularity among the people and they were treated as guardian deities, semidivine beings and attendants of the higher deities. In Jainism we find a fully developed pantheon of Yakṣas and Yakṣis as attendants to the Tīrthaṅkaras. In Buddhism too they had subordinate position as seen on the gateways of the *Stūpas* of Bharhut and Sānchi. In this way the cult of Yakṣas and Yakṣis came into vogue and it became associated with both Buddhism and Jainism.

The images of Yakṣa are considered to be the earliest stone anthropomorphic image in India excepting those from Harappan culture. The image of Pārkham Yakṣa of circa 2nd century B.C. is considered to be the earliest image in human form. Apart from this there are also the traces of Yakṣa images of this period from Mathura.[119]

The images of the Brāhmanical divinities, Jain Tīrthaṅkaras and the Buddha/Bodhisattvas came later into existence respectively.[120]

It seems that when the images of Yakṣa gained popularity among the masses, the Brāhmanical theologians might have felt danger of their faith and consequently they sanctioned the making of the images of the Brāhmanical divinities in human form to maintain their position and when the images of the Brāhmanical divinities became

---

116. Quoted from R.C.Sharma, *Buddhist art of Mathura..*, p.159.
117. *Gītā*, 17.4.
118. R.C.Sharma, *Mathura Museum and Art*, 1976, Mathura,p.28.
119. R.C.Sharma, *Buddhist art of Mathura...*, p. 155.
120. D.L.Snellgrove (Ed.), *Op. cit.*, p.48.

more popular among the people then both Jains and Buddhists also started making the images of the Tirthankaras and the Buddha/Bodhisattvas respectirely in anthropomorphic form may be for the same reason.

(4) One of the major factors which contributed to the emergence of the images of the Buddha was the material prosperity of the period. It is quite obvious that no school of art can flourish and survive without economic support and the people can invest their money for this purpose only in surplus condition. The period of 1st century B.C.-A.D., in which the images of the Buddha are supposed to have been emerged, is considered to be economically prosperous one.

Agriculture appears to have undergone a significant development under the Kuṣāṇas which was certainly the base of the rapid urbanization of the period. The period also witnessed a rapid growth of trade activities both internal and external, the urban centres playing the role of trading centres. The internal trade routes had been existent since the time of the Buddha. There existed the Royal Highway from Pāṭaliputra to Taxila in the North-west, and to the mouth of the Ganges in the east under the Mauryas.[121] "In upper India, Srāvasti (Sahet Mahet), Campā (Bhagalpur), Pāṭaliputra (Patna), Vārānasi, Kausambi, Ayodhyā, Mathura, Sākala and Vidisa, and Ujjayini in the Mālwā region were important centres of trade and it seems that they were interconnected with the main cross-country routes and there were numerous ancillary feeder routes between these towns".[122] Increasing internal trade is indicated by the use of coppers on large scale.[123] Indians mainly exported precious woods, cotton textiles, silk, iron objects and pepper etc. In return they impornted raw materials like crude glass, lead, tin, red coral and asbestos and finished products including wine, and Roman storage and table ware.[124]

The Kuṣāṇa rulers, from the beginning , encouraged the trading activities. Vima Kadphises is said to have started trading activities

---

121. Romila Thapar, *Aśoka and the decline of the Maurya*. New Delhi,1980(reprint), pp.81-82.

122. G.L.Adhya, *Early Indian Economics*, Bombay,1960,p.42.

123. *Ibid*, p.103.

124. Michael Lowe, "Aspect of world trade in the first seven centuries of the Christian Era"' *JRAS*, No. 2, 1971,p.176.

with the Roman empire.[125] The king also started issuing gold coins specially for trade and commerce. It is because of this fulfilment of trade demand that these coins became very popular and captured the whole market.[126] The Kuṣāṇas appear to have brought most of the trade routes under their command.

Rapid growth of trade and urbanization gave impetus to the development of craft and industry. Two Buddhist texts refer to various types of craftsmen : the *Mahāvastu* gives a list of thirty-six kinds of workers in Rājagaha, and the *Milindapañha* mentions seventy-five occupations, mostly artisans.[127] These artisans and craftmen appear to have formed guilds of their own trades, as the practice had been there from earlier times. These guilds held an important position not only in the economic but also in the political life.[128]

The technological development, rapidly developing internal and external trading activities and the growth in agriculture production proves that the soceity under the Kuṣāṇs enjoyed prosperous economic condition.

The Kuṣāṇas extended their royal patronage to the cause of Buddhism. Under the rule of Kaniṣka Buddhism received a positive encouragement from the royal family. The epigraphical evidences show that even some of the feudal kings under the Kuṣāṇas, known as Mahākṣatrapa, Kṣatrapa, Sāhi and Rājā adopted Buddhism and gave their patronage to the faith.[129]

The epigraphical sources also tell us that some of the trader class and general people also gave their patronage to Buddhism.

As stated earlier, the period first century B.C. - A.D. witnessed an uprise of *Bhakti* movement which brought almost all existing Indian religions under its fold. Subsequently, religion became more devotional and it resulted in a greater demand for the images of gods and

---

125. G.L.Adhya, *Op.Cit.* ,p.131.

126. A.Dani & F. Khan, "Kushan Civilization in Pakistan"' in *Central Asia in the Kushan period*. Vol. I,Moscow, 1974,p.102.

127. *Mahāvastu*, iii, 442-3; Milindapañha, p.331, quoted from R.S.Sharma, '*Śudras in Ancient India*, Delhi(2nd ed.) 1980, p.261.

128. G.L.Adhya. *Op.Cit.*, p.84.

129. S. Konow, *Kharoṣṭhi Inscriptions*, Vol. II, part-I, No.XV, Varanasi, 1969, pp.48-49, 23-29, 174.

goddesses for domestic worship not only from the newly rising Brāhmanical cults of Śaivism, Vaiṣnavism etc., but also from Jainism and Buddhism as well. The economically prosperous society made their donations for the making of the images of gods and goddesses. The images of Balarām, Viṣnu, Kārtikaya etc. both in stone and terracotta have been discovered from Mathura and other parts of India belonging to this period and earlier.[130] A large number of Jaina *Āyagapaṭa* and Buddha/Bodhisattva images have been discovered from Mathura which belong to the Kuṣāṇa period.[131] So, the economic prosperity of the period was one of the important factors to the emergence of the Buddha image because when the Buddhists felt the need for the anthropomorphic representation of their Master, they came forward with donations for this noble task as they could afford to do so.

How the affluent economy of the Kuṣāṇa period contributed to the arts and crafts: R.S. Sharma, aptly says, "This would suggest that perhaps in no other period had money economy penetrated so deeply into the life of the common people of towns and suburbs as during this period, a development which fits well with the growth of arts and crafts and the country's flourishing trade with the Roman empire".[132]

Thus, we may sum up, that all the above-mentioned factors equally contributed to the emergence of the Buddha image.

1.7   Priority of places where the first image or images of the Buddha/Bodhisattva evolved - Mathura, Gandhāra or Udyāna- Kashmir.?

As stated earlier, the problem where the first image or images of the Buddha were evolved is yet to be solved. Scholars are divided into three camps : first favouring the Gandhāra origin while the second supporting the Mathura origin and the third theory which was carved by A.K. Narain is that the first image of the

---

130. Devangana Desai, "Social Background of Ancient Indian Terracotta (Cir. 600 B.C. - A.D.600)" in D.P. Chattopadhyaya (Ed.), *History and Society*, Calcutta, 1978, pp. 159-60.

131. R.C. Sharma, *Op. Cit.*, pp. 154-166.

132. R.S. Sharma, *Light on Early Indian Society and Economy*, Bombay, 1966, p. 78.

Buddha/Bodhisattva was carved neither in Gandhāra nor in Mathura but in Udyāna - Kashmir. Let us survey these theories one by one:

1.   Theory of the Gandhāra origin of the Buddha image:

India had developed her contacts with foreign countries several centuries before Christ and it was but natural that the region of Gandhāra was a meeting place of different races. During the Mauryan rule in third and second centuries B.C., contact with foreign countries was further established. The Post-Mauryan period witnessed a series of foreign invasions; most of them were from the Central Asia. All the foreign races, Indo-Greeks, Śakas, Pārthians, Kuṣāṇas - settled down in the North-west region of India and some of them extended their rule in other parts of India. Those who settled in the North-west region, developed their own culture which was the synthesis of all the racial cultures including Indian but Greek cultural current was more dominant because it was more developed than the rest. The style of the Gandhāra school is known to have imitated mainly from the art which had a very long tradition of plastic art and its classical age was in the 5th-4th centuries B.C. So, it was easier for the sculptors of Gandhāra to carve the images of the Buddha than the sculptors of Mathura. They were familiar with the working on stone having a long tradition of image-making of deities. The sculptors carved the images of the Buddha on the model of Apollo, a Greek god as is evident by the Gandhāran Buddha images.

The numismatic evidence is also said to have suggested the Gandhāran theory of the Buddha image. The cross-legged seated human figure on a class of coins of Maues is considered to be the earliest depiction of the Buddha by several scholars. A series of coins of Kadphises I depict the seated human figure which is also considered to be the figure of the Buddha. The coins of Kaniṣka show standing and seated figures of the Buddha with Greek legend BODDO, SAKAMANO BOUDO etc. on it. The reliquary of the time of Kaniṣka discovered at Shāhje-ki-dheri also bears a figure of the Buddha.

A Kharoṣṭhi inscription of the time of Kadaphises I mentions about a Buddha image. The inscription is engraved on a gold plate, states that the body relics of the Bhagavant i.e. the Buddha to be depicted were painted to the life.[133] Further discussion on this inscription will be made in a chapter on Gandhāra.

The reliquary at Bimaran (Gandhāra region) is forwarded as a strong evidence in support of the Gandhāran origin of the Buddha image. This gold relic casket was discovered in relic chamber of a *Stūpa* at Bimaran. The casket is decorated with a series of arched niches with figures of the Buddha flanked by Indra and Brahmā. The pillars of arches bear an oblong cut.[134] A few coins of Azes II were found with the casket and scholars have dated it in about 50 B.C. and if it is so, then this has to be considered as the earliest representation of the Buddha in human form.

The existence of pre-Kaniṣka Buddha image in the Gandhāran region is further supported by the discovery of a number of sculptured panels illustrating, among other things, the figures of the Buddha in a group of monuments at Butkara (Pakistan). D. Faccenna has dated them, mainly on numismatic ground, to the period from the late first century B.C. to early first century A.D.[135]

Let us examine the above mentioned points one by one.

The sculptors who carved the images of the Buddha in Gandhāra may not be treated as foreigners since they were settled down in the North-western region of India in a quite early period. It is true that Greek elements were dominant in the Gandhāran art but when this school emerged in the first century B.C. it had acquired Greaco-Roman characters and also incorporated the Iranian and Indian elements and motifs.[136] The imitation of Hellenistic or Roman trend does not support the view that the first image of the Buddha was carved in Gandhāran school. Regarding this, Coomaraswamy rightly says that sculptures of Gandhāra school may date from first century B.C. probably before Kaniṣka and it attained greatest expansion in Kaniṣka's reign.[137]

---

133. H.W. Bailey *Op. Cit.*

134. D.L. Snellgrove, *Op. Cit.*, pl. 36.

135. D. Faccena, *Op. Cit.*

136. A.K. Coomaraswamy, *History of Indian and Indonesian Art*, New Delhi, 1972, p. 52.

137. *Ibid.*

Though the Gandhāran school was essentially a Hellenistic one but it followed the Indian tradition in its iconography.[138] No doubt the images of the Buddha from Gandhāra inherit the ideals of Apollo but it does not prove that the Gandhāran school produced the first image of the Budha.

As regards the cross-legged human figure on a class of coins of Maues which is identified by some scholars as the figure of the Buddha, will be discussed in the Udyāna-Kashmir section. But majority of scholars do not accept it as the figure of the Buddha. So far as the coins of Kadphises I are concerned, it is also not beyond controversy. V.A. Smith,[139] John Marshall,[140] and Whitehead[141] accept it as a figure of the Buddha. But Coomaraswamy contradicts this identification casting doubt on some raised object of the seated figure. The seating pose is exactly similar to the early images of the Buddha from Mathura except the raised object and its overall appearance is that of a king. Coomaraswamy opined that the raised object held by the right hand of the figure is some hammer-like object, perhaps a scepter; the left hand rests on the thigh, and the elbow is extended, the breadth of the shoulders and slenderness of the waist are conspicuous. Thus, it seems that this personage represent a king and not a Buddha.[142] Coomaraswamy, however, admits that apart from the object held in the hand of the figure, there is similarity of this figure with the early Mathura Buddha images and of the Buddha at Amarāvati.[143] But he has doubt with regard to the identification of the Buddha figure on the coins of Kadaphises. Whether Kadaphises type coins depict the images of the Buddha or not, one can not say precisely because the figure on the coins is not very much clear, but the Kharoṣṭhi inscription of the period of Kujul Kadaphises clearly mentions about the image of the Buddha. However, it also does not

---

138. *Ibid.*

139. V.A. Smith, "Numismatic Notes and Novellies", Part II, *JASB*, 1897, quoted from R.C. Sharma, *Op. Cit.*, p. 151.

140. J. Marshall, "Excavation at Taxila", *ASR*, 1913-14, p. 44, pl. XL.

141. R.B. Whitehead, *Punjab Museum Coins Catelogue*, Oxford, 1914, pp. 181-82.

142. Quoted from R.C. Sharma, *Op. Cit.*, pp. 151-152.

143. *Ibid.*

support the view that the first image of the Buddha was made in the Gandhāran region.

Even the Bimaran reliquary which depicts the figure of the Buddha does not support the view of the Gandhāra origin of the Buddha image because the dating of the reliquary has been challenged by many scholars on various grounds. Coomaraswamy points out that the "methods of excavation nearly ninety years ago were not by any means as critical as they are now, coins in any case merely provide a terminus *past quem*".[144] Lohuizen-de-Leeuw[145] says that the little oblong cut on the pilasters is a late motif and the oldest Gandhāran specimens do not display this shape. On the stylistic ground the author place it in the later half of first century A.D. The author further says that the numismatic evidence tells that the mound of Sirkap near Taxila was in habitation before Kaniṣka because coins have been found only upto the period of Vima Kadaphises. Lohuizen rightly asks that if the Bimaran reliquary is to be dated from about 50 B.C. then why do we not get any Buddha image till Kujula Kadaphises period? [146] Joe Cribb says that there is similarity between the figure of the Buddha depicted on the Bimaran reliquary and on the Kaniṣka reliquary.[147] The other factor favouring the early dating of the reliquary is the four bronze coins discovered with it. Joe Cribb[148] further says that none of the coins were found within the reliquary or the pot containing it, so there is no question of any chronological relationship between them. There is possibility that the *Stūpa's* relic chamber was closed later than the reign of Azes II. According to Cribb, "the coins bear the name of Azes, but have inscriptions, types and control marks, not found on the regular issues of either Azes I or II. Coins of this type could have been issued from the end of the reign of Azes II up to the beginning of Vima Kadphises".[149]

---

144. A.K. Coomaraswamy, *Origin of the Buddha Image...*, p. 3.

145. J.E. Van Lohuizen-de--leeuw, *The Scythian Period*, Leiden, 1949, pp. 84-85.

146. *Ibid.* p. 87.

147. Joe Cribb, "A Re-examination of the Buddha Images on the coins of Kaniṣka ...", in A.K. Narain (Ed.), *Studies in Buddhist Art of South Asia*, Delhi, 1985, p.82.

148. *Ibid.*

149. *ibid.*

There is Kharoṣṭhi inscription depicted on the steatite vase of the casket. But it also does not support the early dating of the reliquary. Sten konow dated it to the latter part of the first century B.C. while Thomas favoured a date between 50 A.D. and 78 A.D.[150] K.K. Dasgupta considers it a product of pre-Kaniṣka period possibly sometime in the middle of first century A.D.[151] Joe Cribb says that the epigraphy is typical of the Kuṣāṇa period, it can not be precisely dated. [152] Thus, on the basis of above mentioned scholar's opinions we may place it to the period around middle of first century A.D.

The images of the Buddha discovered from Butkara (Pakistan) and which have been dated to the period first century B.C. - A.D. may be considered as the earliest representation of the Master from Gandhāran side.

2.   Theory of the Mathura origin of the Buddha Image.

Several factors have been put forth in support of the Mathura origin of the Buddha image, which are as follows:

(A)   As stated earlier that under the impact of Brāhmanic *Bhakti* movement images of different gods and goddesses came into existence which also affected Buddhism and under their influence the images of Buddha/Bodhisattva emerged. Mathura was the chief centre of the *Bhakti* movement. There was no such type of thing in the Gandhāran region.

(B)   Buddhism spread in Mathura in quite early period whereas Gandhāra received Buddhism much later, at least after Mathura. The tradition of image-making started in Mathura from the Mauryan period as testified by the discoveries of Yakṣa statues. A colossal Yakṣa image has been discovered from Pārkham and a Yakṣi statue from Nagla Jhinga[153] belonging to the Mauryan period(Pl. 1A.). The process of art activity in Mathura grew in momentary in the 2nd-1st centuries B.C. as evidenced by the discovery of numerous images of Yakṣa, Yakṣi, Nāga, Balarāma

---

150. R.C. Sharma, *Op, Cit.*, p. 154.

151. K.K. Dasgupta, *Op. Cit.*, p. 151.

152. Joe Cribb, *Op. Cit.*, p. 82.

153. R.C. Sharma, *Op. Cit.*, pp. 154-55.

etc.[154]   Thus, the Mathura school was fully capable to carve the images of the Buddha and due to the canonical interdiction they (the sculptors of Mathura) could not do so earlier.

(C)   One of the main characteristics of the Mathura school of art is that it produces both symbolical and human images of the Buddha. Thus, the Mathura school shows the different stages of evolution of the figure and these stages reflect the gradual development and change in the mental outlook of the society and the chain of events which resulted in the outcome of the Buddha image.[155]

(D)   The three earliest images of the Buddha/Bodhisattva, although carved in the Mathura style, were installed at distant places. One discovered at Kausambi, belongs to the 2nd regnal year of Kaniṣka, the other in the Sārnāth Museum is dated in the 3rd regnal year of Kaniṣka. Another Buddha image at Katrā(Mathura) is supposed to belong to an early date.All these images were from Mathura.[156] These three images are finely executed and show a fairly developed stage of Buddhist art and iconography when the sculptor already had a set model before him. So, the sculptors who carved these images would be expected to posses a long experience of shaping the Buddha figure. The higher artistic tradition of the Buddha images in the early Kaniṣka period prove that Mathura had the tradition of the Buddha image-making before the Kaniṣka period.[157] and it must have aquired a prestige and henceforth exported to other places.[158]

(E)   As stated earlier, there was a tradiition of Yakṣa image- making in Mathura from the time of the Mauryan and Śuṅgan rule. A close examination of Yakṣa and early standing images of the Buddha/Bodhisatttva from Mathura school reflect a good deal of resemblence between the two.[159] Sometimes similarity

---

154. *Ibid.*, p. 155.
155. R.C. Sharma, *Op. Cit.*, pp. 154-55.
156. *Ibid*, pp.155-156.
157. *Ibid.*, p. 156.
158. A.K.Coomaraswamy, *HIIA*, p. 59.
159. R.C.Sharma, *Op.Cit.*, pp. 156-157.

between these two images was so striking that it was very difficult to distinguish between the two and only the inscription made possible the identification, otherwise it led to confusion.[160] Even Sujātā in the *Nidāna Kathā* took the Buddha to be a Yakṣa. Not only the iconography but the aim of installing was at times the same as of a protecting deity.[161] Hence, on the basis of similarity between these two images, Coomaraswamy rightly says that the standing image of the Buddha evolved from the Yakṣa images.[162] Even the elements of later anthropomorphic iconography like nimbus, *Uṣṇiṣa*, were already present in early Indian art before the emergence of the Buddha image.[163] Lohuizen, however , suggests that the early Buddha images derived their ideals from the King- type Mathura images and the two have close affinity.[164] It seems that both Yakṣa and king-type images of Mathura school influenced the early standing images of the Buddha/Bodhisattva but the earlier had greater impact.

So far as the seated images are concerned, there is no such type of Yakṣa image pertaining to the Mathura school belonging to the early period. So, the image of Yakṣa could not be said to have been the prototype of seated Buddha/Bodhisattva images.[165] The earliest seated Buddha images are placed in the beginnning of Kaniṣka's reign and from Kaṭrā and Anyor, does not show resemblence with the Yakṣa images. Coomaraswamy opined that the seated images of the Buddha have its prototype in Yogi-like figures in some Bharhut reliefs as well as similar models found on a few coins of Maues and Kujula Kadaphises and also on some early coins hailing from Ujjaini.[166] The earliest seated figure in early Indian art, barring those of *Paśupati* type figure of the Indus culture, has been found in the sculptured panels discoverd in Bharhut. The figures are carved on the railing of

---

160. *Ibid.*, p. 157.

161. A.K.Coomaraswamy, *Op.Cit.* , p. 18.

162. *Ibid.*, pp.17- 18.

163. *Ibid.*,pp. 15-27.

164. J.E.Van Lohuizen, *Op.Cit.* , pp. 153-160.

165. *Ibid.*, p.154.

166. *Ibid.*, pp. 33-38.

the *Stūpa*, which are now preserved in Indian Museum Calcutta. An architectural fragment of the *Stūpa* shows two seated male figures; both of them wear turbans and ornaments and neither can be identified as ascetics but they are seated in cross-legged meditating pose. The central band of another frieze, depicts a seated ascefic in *Vyākhyāna mudrā*.[167] By observing this figure, R.C.Sharma says that there is a close resemblence in this scene with the event of the *Dharmacakra pravartana* by the Master to his first five disciples at Sārnāth.[168] Another source of inspiration might have been the seated Jain figures depicted on the *Āyagapaṭas*, excavated at Mathura belonging to later half of the first century B.C.[169] i.e. the transitional phase when symbols and image-worship flourished side by side.[170]

Let us survey the features of some of the Buddha/Bodhisattva images of pre-Kaniṣka period which is said to have derived their ideals from the images of Yakṣa, king-type figure and from the Jain *Āyagapaṭa* figures.

(1)   A Bodhisattva image housed in the Lucknow Museum(B.12B) is considered to be an important specimen.[171] The treatment of drapery, the ornaments, the posture of hands, the girdle, modelling of the body etc. show that the image is directly drawn from the early Yakṣa tradition. Between the shoulder and the raised hand in the *Abhaya mudrā*, is the continuation of the *Chauri* which the Yakṣa used to hold in hand. The 'U' shaped necklace also supports its early origin.

(2)   A second figure which is also dated to the pre-Kaniṣka period is housed in the Lucknow Museum. The figure resembles the Yakṣa tradition with heavy body, frontal pose and especially the bulging scarf generally found in early sculptures.[172] On the basis of its provenance(Kaṅkāli mound) it was previously confused to be a Jaina statue but absence of *Śrivatsa* mark on

---

167. R.C.Sharma, *Op,Cit.*, pp. 161-162.

168. *Ibid.*, p. 162.

169. *Ibid.*, p. 162.

170. *Ibid.*

171. *Ibid.*, p. 162.

172. *Ibid.*, 159.

the chest and clothes on the body, while Jaina figure should
be represented nude, its features decifer it as a Buddha, was
derived from the king-type of images from Mathura.[173]

(3)   An image which shows the resemblence with the Jaina *Āyagapaṭa*
        figure, has been found in a fragmentary post in red sandstone,
        housed in Mathura Museum (No.H.12). Earlier it was also
        wrongly identified as a Jaina figure. [174] In this figure the Buddha
        is shown as an ascetic flanked by four *Lokapālas* holding alms
        bowl in their hands. The Buddha sits on a high pedestal with
        his right hand in *Abhayamudrā* and the left hand is placed on
        the left thigh. Below the pedestal, a pair of lions seated back
        to back, are shown. It has been dated to the end of 1st century
        B.C.

(4)   Another pre-Kaniṣka Budddha image is housed in the Museum
        fuer Voelkerkunde, Muenich(fig. 81). It represents the Buddha
        in the Kaṭrā style with shaven head, prominent top knot, *Ūrṇā*
        between the eyebrows, small earlobes, right hand raised in
        Abhaya mudrā. The left hand is broken.[175]

(5)   Next representative of pre-Kaniṣka Buddha image, is a frag-
        mentary sculpture discovered at Galteswar in Mathura, housed
        in Mathura Museum.[176] This is a Bodhisattva image as the
        inscription says. Only the left leg,right foot and left hand of
        the Bodhisattva along with the left foot of an attendant on the
        side are extant. The central part of the pedestal bears a rampant
        winged lion and a female worshipper. The fragmentary
        evidence proves that it was carved in the style similar to the
        Kaṭrā Boddhisattva image. The most important thing of the
        sculpture is that it is incised with a four lines of inscription
        which informs us that a female worshipper had set up this
        Bodhisattva image and it was dedicated to the Sarvāstivādin
        monks for their well being. Equally important is the reference
        to some Kṣatrapas in the inscription which is of vital importance

---

173. J.E.Van Lohuizen, *Op. Cit.*, p. 155-160.
174. R.C.Sharma, *Op.Cit.*, p.163.
175. *Ibid.*, p. 164.
176. *Ibid.*

in deciding the age of this piece. Though, scholars are not unanimous about the reign of Kṣatrapa, this piece of sculpture has been dated to the pre-Kaniṣka period.

In this way, we find that Mathura school has a tradition of Buddha/Bodhisattva image making from the pre-Kaniṣka period.

The early Ujjain Coins which represent the seated human figure which according to Coomaraswamy, is the Buddha figure. The image is seated on a lotus seat and the *Bodhi* tree is seen to the right of the image. Though the date of the coin is not certain but it is assigned to the 1st century A.D.[177]

3.   Theory of the Udayāna-Kashmir Origin of the Buddha /Bodhisattva Image:

Theory of the Udayāna-Kashmir Origin of the Buddha/Bodhisattva image is put forth by A.K.Narain.

A.K.Narain has based his argument on the early expansion of Sarvāstivādins school in the Udyāna-Kashmir region and the controversial coin of Śaka King Maues which depicts a cross-legged human figure, which according to Narain, is a figure of the Buddha or Bodhisattva. He credits the first human representation of Buddha/Bodhisattva to the Śaka- Sarvāstivāda combination in the time of Maues(95-75 B.C.) in the north of Gandhāra to the Swāt Valley and Kashmir.[178]

Let us survey the theory:

(A)   The Śakas who captured a much larger part of India came to India after the Greeks. Earlier they (the Śakas) moved to

South from the upper Ili regions by crossing Pamirs and spread out in the submountain Karakorum and finally reached Chipin i.e. the Swāt valley and adjoining areas of Kashmir.[179] By about 100 B.C. the Śakas under the leadership of their chief Maues had controlled a considerable portion of North and North-east Gandhāra. But before reaching the Swāt valley, the Śakas had

---

177. A.K.Coomaraswamy, *Origin of the Buddha Image...*, p.16.

178. A.K.Narain,   "First Images of the Buddha and Bodhisattva: Ideology and Chronology" in *Studies in Buddhist Art of South Asia...*pp. 1-16.

179. A.K. Narain, *The Indo-Greeks*, Oxford University Press, 1980, Chapter VI.

frquented the routes and regions of Karakorum and provided political control and protection which paved the way for traders and missionaries alike. All this in turn led to much artistic and religious activities. The epigraphic material engraved on the Karakorum rocks dates from the times of Maues[180] and some of these epigraphs are associated with the sketches and motifs. The carvings not only indicate the aniconic and iconic phases of religious art but also how the transition from one to other might have taken place of particualr importance is the Rock Carving No. 18 of Chilas II where the seated figure, according to Narain, is surely that of the Buddha/Bodhisattva. The inscription below the figure has been read as *Budhaotasa*, Narain suggests that perhaps it stands for Bodhisattva.[181]

(B)   Narain also says that the Śakas needed to strengthen the legitimizition of their rule and the Sarvāstivādins who were very prominent in the Kashmir region, needed royal patronage for their ideology of realism and the Śakas were controlling the northern routes to and from central Asia and their protection must have been of great help to the Buddhist missionaries as well as to the traders, most of whom were probably Buddhists. The Śakas and the Sarvāstivādins joined hands which led to the emergence of the first image of the Buddha and Bodhisattva. The epigraphical sources prove that the Śakas were patron of Sarvāstivādins.[182]

(C)   The cross-legged seated figure on a class of coins of Maues, as stated above, is considered by A.K. Narain as the earliest depiction of the Buddha/Bodhisattva in human form. The figure is shown seated on a throne or pedestal or a cushion with both hands, according to Narain, meeting on the lap as if on Meditation.[183] There is monogram in the right field and below the monogram there is a small horizontal line placed between the elbow point and the knee. This has been identified as a

---

180. A.H. Dani, *Chilas, the city of Nangā Parvat (Dyamar)*, Islamabad, 1983, pp. 91-128.
181. A.K. Narain, *OP. Cit.*, p. 5.
182. *Ibid.*, p. 6-7.
183. *Ibid.*, p. 7.

sword or accross, his knees. But Narain says that it is not a sword on or a part of it, actually some of the specimens seems to link the tiny horizontal line with the monogam or with the space occupied by the monogram in such a manner that it looks like a part of it.[184]

As regards the identification of this cross-legged seated figure as the Buddha/Bodhisattva, scholars are not unanimous. Whitehead described it as king seated on a raised cushion.[185] V.A. Smith identified the figure as a deity or a king.[186] Coomaraswamy does not favour the identification of the figure as the Buddha and says that though, the two hands are folded in the lap, but there is a horizontal bar extended to the right which may be a sword or sceptre or possibly the back edge of a throne or seat.[187] But Longworth Dames considered it as the Buddha figure.[188] Narain, as stated above, also identifies it as the Buddha figure and on the basis of this coin type figure he says that the first images of the Buddha/Bodhisattva emerged neither in Gandhāra nor Mathura but in the territories north of Gandhāra i.e. Udyāna-Kashmir, from where Maues first ruled before entering into the Taxila.[189]

CONCLUSION :

After surveying the different theories of the origin of the Buddha image it reveals that the anthropomorphic images of the Buddha were known at least before the Kaniṣka period.

In Gandhāra region the images of the Buddha were known from the Kujula Kadaphises period as evidenced by a Kharoṣthi inscription. The discoveries of a number of sculptured panels, displaying, *inter alia*, the images of the Buddha in a group of monuments discovered at Butkara (Pakistan) have been dated to the period from late first century B.C. to early first century A.D.[190] The

---

184. *Ibid*.

185. R.B. Whitehead *Punjab Museum coins Catalogue*, Oxford, 1941, p. 102.

186. V.A. Smith, *Catalogue of Coins in the Indian Museum*, Calcutta, 1906, p. 40, pl.VIII,4.

187. A.K. Coomaraswamy, *Op. Cit.*, p. 16.

188. *JRABA*, 1914, p. 793.

189. A.K. Narain, *Op. Cit.*, p. 10.

190. D.Faccena, *Op. Cit.*, pp. 150ff.

reliquary of Bimaran is not so early as considered by some scholars. The theory of Foucher that the first image of the Buddha was carved in the Gandhāran region, no longer stands.

The Mathura school also carved the images of the Buddha in the same period (i.e. between 1st century B.C. and 1st century A.D.), but named it Bodhisattva due to canonical interdiction.[191]

The theory of A.K. Narain that the first images of Buddha/Bodhisattva were made in the Udyāna-Kashmir region is a new addition but there are some reservations in accepting this theory.

(1)   Whether the region Udyāna-Kashmir can be considered as a separate part from Gandhāra because most of the works mention Udyāna-Kashmir as a part of Gandhāra. Even in Buddhist literature both Gandhāra and Kashmir are mentioned as one region.

(2)   No doubt the Śakas patronised Buddhism, but the coins of Maues, which depicts a cross-legged human figure, can not be identified as Buddha/Bodhisattva with certainty as the figure is very much rubbed. Buddhist images discovered in Kashmir mostly belong to the early medieval period. Thus, Narain's hypothesis can not be considered, i.e., that first images of the Buddha/Bodhisattva were carved in the Kashmir region.

Thus, the choice lies on between Gandhāra and Mathura. Whether the first image of the Buddha/Bodhisattva emerged in Gandhāra or in Mathura, it also can not be stated precisely due to lack of adequate material. The stylistic features of both the schools reveal that both of them developed independently and most probably both Gandhāra and Mathura school produced the images of the Buddha/Bodhisattva simultaneously in the 1st century B.C. - A.D.

---

191. K. K. Dasgupta, *Op. Cit.*, p.151

CHAPTER - 2

ICONOGRAPHY OF THE BUDDHA IMAGES IN MATHURA SCHOOL OF ART

2.1 : MATHURA : Historical perspective of the place

2.2 : Buddhism in Mathura

2.3 : Mathura school of Buddhist art.

2.4 : Iconography of the Buddha images.

2.1 Mathura : Historical perspective of the place:

Mathura is situated on the right bank of the Yamuna at a distance of 58 km. to North-west of Agra.[1] All major early Indian religions i.e. Buddhism, Jainism and Brāhmanism flourished at Mathura. It was also a famous centre of art.

Apart from its religious background, other important factors which led Mathura to rise to the eminence, was its ideal geographical position. It was situated on the junction point of many important land routes as well as a major river. Several roads from Mathura which connected the two important high ways of ancient India viz. Uttarāpatha (the Northern High way) and Dakṣināpatha (the Southern High way).[2] The Uttarāpatha was an important trade route which connected Mathura with Northern and North-west regions of India via Taxila, Peshawar, Kabul, upto Bactria. The Dakṣināpath joined Mathura with central India particularly Mālwā and Deccan plateau,

---

1. R.C. Sharma, *Buddhist Art of Mathura*, Delhi, 19, p.3.
2. S.G. Bajpai, "Trade Routes, Commerce and communications patterns from the Post Mauryan to the Kuṣāna period," (paper presented in the *International Seminar on Mathura* held in New Delhi, 1980).

upto the Western sea coast. Through these trade routes Mathura had also linkage with several notable capital cities such as Ujjayini, Prastisṭhān[3] etc. The Royal Highway from Pāṭaliputra to Taxila was built under the Mauryan rule,[4] which also connect Mathura.

Mathura was one of the important urban centres among Śrāvasti, Vidisā, Vārānasi, and Campā etc. These cities were inter-connected with main cross-country routes and numerous other ancillary routes between these towns.[5] Besides being well connected with land routes, Mathura was also an important centre of navigation and hence maritime trade also developed there as attested by literary sources.[6]

The soil of Mathura is generally yellow which is not much fertile. That is why it is said that the land around Mathura not being fertile, became a big city and an important centre of trade and commerce, mainly due to its strategic geographical[7] location. So, its ideal geographical situation and the network of roads and tracks transformed it into a meeting place of several cultural currents through traders, political aspirants and religious preachers and followers[8] of different religions. With the invasion of foreign races from 2nd century B.C. onwards and their settlement in North-west and North India, Mathura also came under their influence. The period from the 2nd century B.C. to the Kuṣāṇa period, saw the rise of mercantile community and strengthening of guild system all over India;[9] many trades, crafts, occupational groups participated and generated a rich material culture which is attested by the variety and huge quantity of coins as well as massive urbanisation.[10]

Mathura became a great centre of trade and craft and of religion and administration in the first three centuries of the Christian era when it was under the stronghold of the Śaka and the Kuṣāṇa power.[11]

---

3.   *Ibid.*
4.   D.N. Jha, *Ancient India, An Introductory Outline*, Delhi, 1986, p.80.
5.   G.L. Adhya, *Early Indian Economics*, Bombay, 1960, p. 97.
6.   R.C. Sharma, *Op. cit.*, pp. 5-6.
7.   R.S. Sharma, *Perspective in Social and Economic History of India*, Delhi, 1983, 'Trends in Economic History of Mathura (c.300 B.C. - A.D. 300)"
8.   R.C. Sharma, *Op. Cit.*p. 6.
9.   Romila Thapar, *A History of India*, Vol.I, Penguin Books, 1972, p. 109.
10. R.S. Sharma, *Op. Cit.*
11. R.C. Sharma, *Op. cit.*, p.7.

The Soṅkh excavation at Mathura by H. Hartel revealed seven levels of densely built area of houses of the Kuṣāṇa period compared to only one of the Gupta phase,[12] which clearly attests Mathura's prosperity during the Kuṣāṇa period. The learned preceptors, stone cutters, skilled masons, sculptors and other artisans effectively utilized the donated wealth from mercantile community and people from other sections of society which resulted in the creation of numerous icons, religious shrines, monasteries and other monuments etc. in Mathura.

After the disruption of the Kuṣāṇa rule, Mathura lost its political status. The repeated invasions also disintegrated its trade activities.[13] However, development in art and architecture and religion somehow continued at least upto the Gupta period and after that Mathura's fame rested only as a religious place.

2.2　Buddhism in Mathura:

Mathura was the great centre of Buddhism right from the days of the Buddha down to the Imperial Guptas. The *Aṅguttaranikāya*, mentions that the Buddha had visited the place and two of his chief disciples, Mahākaccāna and Revata were associated with Mathura.[14] Mahādeva who was associated with the 2nd Buddhist Council, held at Vaiśhāli, was the son of a Brāhmaṇ of Mathura.[15] Once when the Buddha was going from Mathura to Verañjā, He had halted under a tree by the wayside and a large number of house-holders came and worshipped Him.[16] Upagupta, the teacher of Aśoka was also associated with Mathura. He had also established a monastery called Upagupta monastery in Mathura which had occupied an important place in the history of Buddhism.[17] By the time of third Buddhist council, held at Pāṭaliputra under the patronage of Aśoka, Buddhism was divided into different sects, of which Sarvāstivāda was one. The Sarvāstivādins

---

12. H. Hartel,,"Some Results of the Excavation of Sonkh : A Preliminary Report, "*German Scholars on India*, Vol. II, Bombay, 1976, p. 75.

13. R.C. Sharma, *Op. Cit.*, p. 7.

14. *Aṅquttara Nikaya* (PTS), II, p. 57; P.D. Mittal, *Braj Ke Dharma Sampradāyon Kā Itihās* (Hindi), Delhi, 1968, p.33.

15. K.D. Bajpai, *Mathura*, 1955, p. 14.

16. *Aṅquttara Nikaya*(PTS), *Op. cit.*.

17. B.N. Choudhury, *Buddhist Centres in Ancient India*, Calcutta, 1982, p.40.

held an honourable position in Magadha, but later on having difference with Moggaliputta Tissa, the author of the *Kathāvathu*, and some other reasons, they were compelled to leave Pāṭaliputra. Subsequently they migrated to the north and first reached Mathura where under the leadership of Upagupta they founded a centre of their own school of thought from where later on, it spread to North-west provices of India, Kashmir and Central Asia.[18]

Various donatory inscriptions and other sources tell us about the Sarvāstivādin monks and nuns of the Śaka-Kuṣāna period belonging to Mathura.[19] The condition of Buddhism at Mathura is also known from the accounts of Chinese travellars. Fa-Hien, who visited India in the 5th century A.D., presents a prosperous picture of Buddhism at Mathura. He mentions that there were 20 *Saṅghārāmas* on both sides of the river Yamuna and 300 monks resided in them.[20] The number of monks and priests some time would be as many as ten thousand as stated by the traveller.[21] Apart from these, he also mentions about six *Stūpas* out of which Sāriputra *Stūpa* was most important and venerble.[22] Here it is notable that during the Gupta period Bhāgavatism was in the flourishing condition and was favoured by the Gupta Kings. Mathura was also one of the main centres of Bhāgavatism.

Another Chinese traveller, Hiuen-Tsang who came to India in the 7th century A.D. also mentions same number (20) of Buddhist monasteries at Mathura but the number of Buddhist monks were reduced to 200.[23] He had seen a number of *Stūpas* at Mathura. He further mentions that besides Theravādins, Mahāyāna, and Mahāyānic deities were also worshipped there.[24]

---

18.   A.C. Banerjee, *Sarvāstivāda Literature.*, p. 5.
19.   K.L. Hazra, *Royal Patronage of Buddhism in Ancient India*, Delhi, 1984, pp. 128,147-150,154.
20.   B.N. Choudhury, *Op. Cit.*, p.40.
21.   K.A.N. Shastri, "Chinese travellers", in P.V. Bapat (Ed.), *2500 years of Buddhism*, 1976(reprint), p. 225.
22.   R.C. Sharma, *Op. Cit.*, p. 45.
23.   B.N. Choudhury, *Op. Cit.*, pp. 40-41.
24.   R.C. Sharma, *Op. Cit.*, p.46.

However, we should not fully rely upon his (Hiuen-Tsang) statement as he appeared to have mixed up the description of Mathura with some other places. In fact, during the time of his visit Buddhism was in declining stage as attested by five Brāhminical temples in his account.[25] The Huna invasions had severely damaged the Buddhist establishments of Mathura. Buddhist sculputres after 600 A.D. hardly exist at Mathura. The revival of the Brāhmanical religion was another cause for the disappearence of Buddhism from Mathura. The decline of trade and commerce in Gupta period, the shifting of political power during that time led Buddhism on the path of decline.[26] The final blow perhaps came from the invasions of Mahmud of Ghazani in 1017 A.D. as a result of which the monasteries became dislocated and deserted.

Thus, on the basis of literary traditions and the accounts of Chinese travellers etc., we may conclude that efforts were made to spread the message of the Buddha in Mathura region from the life time of the Buddha upto the Gupta period.

2.3　Mathura school of Buddhist art:

The Mathura school of Buddhist art occupies a prominent place in the history of Indian art. Regarding the origin of this school, though, scholars are not unanimous, however, archaeological evidences suggest that Mathura school of art began sometime in the 3rd or 2nd century B.C.[27] The early art products of this school are closely related to the central Indian school of Bharhut and Sānchi.[28] During the Mauryan-Śungan period the art specimens of Mathura school were Yakṣa-Yakṣi etc. in stone and terracottas, Jain *Āyagapaṭas* and the representation of some other deities. John Marshall has classified pre-Kuṣāṇ art products of Mathura in three groups, the first or earliest group is assignable to the 2nd century B.C., the second group belongs to the first century B.C. and the third one is associated with the rule

---

25.　*Ibid.*, p.47.

26.　*Ibid.*

27.　A.K. Coomaraswamy, *History of Indian and Indonesian Art*, Delhi, 1972, p. 37; V.S. Agrawala, *Studies in Indian Art*, Varanasi, 1965, pp. 114-115.

28.　S.K. Saraswati, *A Survey of Indian Sculpture*, Calcutta, 1957, pp. 52,62.

of local Kṣatrapas.[29] The glorious period of Mathura school of art began with the Cristian era.[30] Mathura school of art created and introduced cult images in early Indian religions, particularly in Brāhmanism, Jainism and Buddhism which gave a new dimention to Indian art.

The early standing image of the Buddha and Bodhisattva discovered at Mathura show a great affinity with the Yakṣa images while seated ones to the Jaina *Āyagapaṭa* figures. So, on the basis of these resemblance it is aptly said that the early standing images of the Buddha were introduced from the prototype of those of the Yakṣa images.[31]

Mathura school of art had established its own ideals and places. The art products of this school are made of the local available materials, spotted red sandstone from Sikri.[32] Stylistically, Mathura school of art developed in different stages.[33]

STAGE -A : The features of the sculptures of this stage are that, the images are heavy and voluminous with better frontal treatment as seen in the old Yakṣa or early Buddha/Bodhisattva images.

STAGE - B : In the sculputres of this stage we find a tendency of reducing the heavy physique by a relaxation of the flesh and an open eyed smiling countenance which characterize the seated and standing Buddha and Bodhisattva image.

STAGE - C : The features of the sculpures of this stage are mundane, sensuous, and graceful. In this stage we find a good deal of foreign features mixed up with indigenous trends. Thus, the sculptures of this stage show the synthesis of indigenous and foreign features. The sculptures of this stage are assigned to the Śaka-Kuṣāṇa period.

STAGE - D : The sculptures belonging to this stage are assignable to the Kuṣāṇo-Gupta period. The features of sculptures show a

---

29.   V.A. Smith (Ed.), *The Cambridge History of India*, Vol.I, Cambridge, 1922, pp. 632-633.

30.   S.K. Saraswati, *Op. Cit.*, p.2.

31.   A.K. Coomaraswamy, *The Origin of the Buddha Image*, 1972, pp. 18-19.

32.   R.C. Majumdar (Ed.), *The Age of Imperial Unity*, Bombay, 1953, p.522.

33.   R.C. Sharma, *Op. Cit.*, pp.132-133; R.C. Majumdar (Ed.), *Op. Cit.*, pp. 522-523.

tendency of growing aversion for the foreign elements and inclination for creating a subtle expression.

STAGE - E : The sculptures of this stage belong to the Gupta period which represent the classical culmination of the sculptures. The art sepcimens are rather detached, supramundane, expressive of inner peace, calm, grace and an indescribable harmony of physical charm and intellectual awakening. Mathura school of art shows decline in the 6th century A.D. due to various causes like the Huna invasions, political turmoil, the rise of other art centres, decline of trade and commerce etc.

Mathura school of art has produced a vareity of forms, open to foreign influence from 2nd century B.C. to 3rd century A.D., it not only expressed indigenous motifs but foreign themes and motifs also like Bachchanalian scene etc.- were created.[34] On the basis of the representation of Greek themes in Mathura school of art, Cunningham, Smith, and Grunwedel, conclude that Bactrian sculptors shaped Mathura school and it owes its art forms to Greek convensions. But Coomaraswamy has refuted this idea and says that Mathura art tradition was native one and the natural flow of Indian art.[35] Later on, Hellenistic and other foreign elements were assimilated in the sculptures of Mathura.[36] The Mathura school transformed the earlier mode of symbolic representation into a new way of anthropomorphic representation. It utilized the new dimension of anthropomorphic representation into depicting deities of various religions such as Brāhmanism, Jainism, and Buddhism. A number of deities evolved and were representated for the first time such as Viṣṇu, Śiva, Kṛiṣṇa, Balarāma, Jina, Buddha and Bodhisattva etc. This eclectic and non-sectarian school of art at Mathura, besides producing the images of the deities of different religions, also produced the deities of other local cults such as Yakṣa-Yakṣi, Nāga-Nāgi, Kubera- Hāriti etc. Apart from these images, the Mathura school of art also produced the images of kings, like Vima Kadaphises and Kaniṣka.[37] Representation of female beauty, divine as well as mortal, is one of the features of Mathura

34. A.K. Coomaraswamy, *History of Indian and Indonesian Art....*, p.63; V.S. Agarwala, *Op. Cit.*, pp. 149-151.

35. A.K. Coomaraswamy, *Op. Cit.*, p.60; R.C. Sharma, *Op. Cit.* pp. 133-34.

36. Stella Kramrisch, *Indian sculpture*, Delhi, 1981, p. 42.

37. A.K. Coomaraswamy, *Op. Cit.*, pp. 66-68; V.S. Agrawala, *Op. Cit.*, pp. 181-188, 189-193.

school, which later on persisted till medieval period.[38] Voluptuous female beauty with ample breasts, heavy hips, graceful postures and attractive looks were carved. These were carved on pillars, slabs and *Āyagapaṭas* which have been found not only in and around Mathura but also upto Sanghol in Punjab.[39] Besides anthropomorphic images, Mathura school employed numerous decorative motifs, animals and composite figures common in Indian religions and also having metaphysical connotation besides being a decorative motif.[40] Apart from the sculptures in spotted red sandstone, Mathura art produced terracotta female figures related to fertility cult which later, along with male figures, developed in Mithuna or Dampati plaques.[41]

2.4   Iconography of the Buddha images

The anthropomorphic images of the Buddha/Bodhisattva emerged in Mathura school of art in about first century B.C.-A.D. [42] The sculptors of Mathura school carved tiny figures of the Buddha /Bodhisattva on *Āyagapaṭas* on the models of Jina whereas the standing figures were carved on the models of Yakṣa. Earlier the sculptors of Mathura school captioning the Buddhist deity as Bodhisattva, not as the Buddha, may be due to the hangover of the old injunction forbidding the representation of the Buddha in human form. This hesitation continued in the early Kuṣāṇa period also.

To trace the important characteristic of the images of the Buddha/Bodhisattva through the ages, the images have been divided into different phases:

The characteristic features of the Pre-Kaniṣka Buddha/Bodhisattva images.

(1)   The standing images of the Buddha/Bodhisattva are influenced by the Yakṣa images.

(2)   Small figures are devoid of the *Uṣniṣa*, but the others are shown with a snail shell (Kapārḍa) tendency.

---

38. A.K. Coomaraswamy, *Op. Cit.*, pp. 56- 62.
39. S.P. Gupta (Ed.), *Kushana Sculptures from Sanghol*, Vol. I, New Delhi, 1985.
40. R.C. Sharma, *Op. Cit.*, p. 138.
41. *Ibid.*, pp. 137-138; R.C. Majumdar (Ed.), *Op. Cit.*, pp. 523-524.
42. *Ibid.*, p. 173.

(3)   The eyes of the images are protruding with a short line at the outer corner.

(4)   The earlobes are not well developed like the later images.

(5)   The chest of the images is well developed.

(6)   The navel is deep.

(7)   The inconspicuous drapery is marked with an incised line suspending from the shoulder in the small figures. The bareness is broken by a few lappete on the left shoulder, but the chest does not indicate any pleat as the cloth is transparent muslin. Sometimes the upper part of the body shows just a bulging scarf touching the left shoulder only.

(8)   The right hand is in *Abhayamudrā* but the left hand is not placed on the knee. The elbow is raised up and does not rest on the lap.

(9)   The thumb of the images is pressed against the fingers.

(10)   The girdle is generally absent.

(11)   The seat generally shows a few tiers, looking like an altar. Sometimes the lions are also seen supporting the seat and there is no place between the two animals.

(12)   The space of the halo is plain and indicated by an incised circle.

(13)   Generally no date or era is inscribed on the images.

(14)   The overall features of the pre-Kaniṣka Buddha/Bodhisattva images are not very refined, they do not show the imprints of the skilled hands and maturity of the artists.[43]

Some of the specimens of images of the pre-Kaniṣka period which bear above-mentioned features are as follows:[44]

(1)   The first representative of this period is a headless Bodhisattva image, discovered at Kaṅkāli mound of Mathura now housed in the State Museum Lucknow (No. B.12b). It is considered as

43. J.E. Van Lohuizen-de-Leeuw,*The Scythian Period*, Leiden, 1949, p. 71; R.C. Sharma, *Op. Cit.*, pp. 173-174.

44. V.S. Agrawala, "Mathura Museum Catalogue", *JUPHS*, Lucknow, 1948, p. 63; J.E. Van Lohuizen, *Op. Cit.*, p. 171.

one of the earliest Bodhisattva images which belongs to 1st century B.C.

(2)   The next representative of this period is a stone slab discovered from the Kaṅkāli mound of Mathura, now housed in the Lucknow Museum (N0. j.531). The slab depicts a scene of discourse between the Buddha and a king. In this figure the body of the Buddha is almost bare except a scarf on the left shoulder. The Buddha is shown here dwarfish like a Yakṣa with a small *Uṣniṣa* over the head. The halo behind the head is devoid of any carving. The slab is dated to 1st century B.C.

(3)   A medallian on the railpost of Mathura, housed in the State Museum Lucknow (No. J.295 or 339), showing a rider on the horseback with a groom in front. It reminds us the scene of *Mahābhiniṣkramaṇa* of Siddhartha, one of the important events of the Master's life. The style of image and double knotted turbans on the head indicates the trend of the Suṅgan period.

(4)   A carved slab discovered at Mathura, housed in the Mathura Museum (N. H.12), depicting the scene of the offering of alms by the four Lokapālas to the Buddha. The bulky body of the Buddha, absense of the halo, turbans of the attending Lokapālas, position and postures of lions etc. are the notable features of the scene.[45]

(5)   A Toraṇ beam discovered at Mathura belongs to the period of 1st century B.C., housed in the Mathura Museum (No. M3), represent a stage of transition between the symbols and the images. The *Dharmacakra* and *Bodhigṛha* on the lions are depicted on the one side of the beam and on the otherside there is a small human figure of the Buddha seated in *Padmāsan*. The right hand of the Buddha is turned in *Abhaya mudrā* and He is being worshipped by Indra and his followers.[46]

(6)   A fragmentary seated Bodhisattva image discovered at Galtesvara near the Kaṭra site of Mathura city, housed in the Mathura Museum (No.A.66) is the next representative of this period.

---

45.   J.E. Lohuizen, *Op. Cit.*, pp. 157-158, fig. 27.
46.   L. Bachhofer, *Early Indian Sculpture*, Vol. II, Delhi, 1973, pp. 27-28.

The sculpture shows Boddhisattva's left leg with the left hand placed on it and the right foot might have been in *Padmāsana*.[47] The pedestal shows a winged lion and a female worshipper in adoration pose. The style of the sculpture represents a developed stage than the rest of this period. The important thing of this sculpture is the epigraph depicted on the pedestal which records that this Bodhisattva image was installed by a female devotee named Nanda for the acceptance of the Sarvāstivādin monks. The other important information which is mentioned in this epigraph is that it records the name of some of the Kṣatrapas.[48] Commenting on this sculpture V.S. Agrawala opined, "the reference to a Kṣatrapa and the Sarvāstivādin monks as well as the early form of writing together with style of carving and spotless hard stone makes it highly probable that the present Bodhisattva image goes back to the time of the earlier Kṣtrapas who ruled at Mathura in the late 1st century B.C."[49]

(7)   Another fragmentary sculpture discovered at Govindnagar, housed in the Mathura Museum (No. 76.104) is identified as the Buddha image belonging to this period. The image shows the upper part of the Buddha with *Uṣnīṣa*. The right hand of the image is turned in *Abhayamudrā*. The clumsy feature illustrates the archaic trends.[50]

The characteristic features of the seated Buddha/Bodhisattva images belonging to the Kaniṣka period.

A special type of change has been in the seated images of the Buddha/Bodhisattva of the Kaniṣka period. Van Lohuizen has termed them as 'canonised or *Karpardin* Buddha'. R.C. Sharma and Van Lohuizen have noticed several distinguishing features of this type of images.[51]

---

47. R.C. Sharma, *Op. Cit.*, p. 175, fig. 78.
48. *Ibid.*, p. 164, fig. 96.
49. V.S. Agrawala, *Op. Cit.*, p. 63.
50. R.C. Sharma, *Op. Cit.*, p. 175.
51. J.E. Van Lohuizen, *Op. Cit.*, p. 150; R.C. Sharma, *Op. Cit.*, pp. 176-78.

The distinguishing features of this type of images are as follows:

(1)   The sculptures of this type are usually carved in the high relief not in round.

(2)   The halo of the sculpture is large and plain in most cases and its edges are scalloped.

(3)   The head of the image is shaven or tied with cloth and the top knot is shaped like snail shell or *Kapārda*. This is why Lohuizen termed these type of images as *Kapardin*.

(4)   The back slab of the sculptures show leaves which indicates the tree.

(5)   The almond shaped eyes are wide open looking straight and not closed as in meditation. There is circular mark or *Ūrṇā* between the eyebrows.

(6)   The earlobes are small.

(7)   The right hand is turned in *Abhayamudrā* and the left hand is resting on the thigh or sometimes clenched which indicate a commanding attitude of a prince or king like a *Cakravartin* style.

(8)   The soles of the feet are marked by auspicious symbols such as *Dharmacakra* or *Triratna* etc.

(9)   The chest portion of the body is prominent and the navel is deep.

(10)   The features of sculptures are clear, expressive and slightly smiling.

(11)   The seat of the images is shaped like an altar with ridges, supported by three or two lions on the pedestal. Some times the central lion is replaced by an object of worship like wheel, *Triratna* etc. and sometimes the pedestal is plain or occupied only by two devotees.

(12)   The garment of the sculptures is generally transparent which covers the left shoulder only and the upper arm shows thick and heavy pleats. The lower garment covers half of the leg only and the edges of the garment fall on the seat.

(13)   Generally inscriptions are engraved on the seat of the sculptures which recorded the date, name of the king or donors etc. and purpose of the installation.

(14)   Generally the deity is flanked by an attendant on each side who usually carries fly-whisk.

(15)   The upper two corners of the slab are occupied by celestials flying in the sky with wreaths in their hands.

(16)   On the whole the sculptures are stiff and straight. The open eyes with heavy physique expresses power and fearlessness towards the world.

Some of the specimen of the seated Buddha/Bodhisattva images of the Kaniṣka period are as follows:

(1)   A Bodhisattva image discovered at Katrā housed in the Mathura Museum (No. A.I) is perhaps the finest piece among the seated images of the Kaniṣka period. The image is shown in *Padmāsana* posture with right hand in *Abhayamudrā* and left hand is placed on the left thigh. The throne of the sculpture is supported by three lions. The body of the Bodhisattva in the sculpture looks stiff and mascular. The transparent robe covers the left shoulder only. The head of the Bodhisattva is covered like a cap, the top of which is termed as a snail shell or *Kapārda*. The eyes are open and the earlobes are long. The plain halo is scalloped at the edge. The Bodhisattva is flanked by two attendants in indigenous style. On the upper corner of the slab there are two more attendants in flying poses. On the edge of the throne an inscription is engraved in Brāhmi script of the Kuṣāṇa period which records that Amohasiya the mother of Buddharakṣita had set up this Bodhisattva image in association with her parents in her own temple for the welfare and happiness of all beings.[52]

(2)   A headless Buddha image discovered at Anyor near Govardhan, housed in the Mathura Museum (No. A.2) is the next representative of this age of figures. Though this image is similar to the above mentioned Katrā image, it differs in some

---

52. R.C. Sharma, *Op. Cit.*, p. 178, fig. 79.

respects. [53] The central lion on the pedestal is missing and left hand is clenched. The back slab and the flying attendants are also missing. Unlike Katrā, the Anyor image inscription records that it is a Buddha image.[54] Vogel considered it as the oldest Buddha image found at Mathura with designation of Buddha in its epigraph.[55]

(3)   A headless and armless seated Bodhisattva image discovered at Sonkh,[56] is housed in Mathura Museum (No. 20.1602). The upper garment of the sculpture is transparent except the few pleats on the left hand, are noticed. The frill of the garment is seen on the pedestal as in the Katrā image. The pedestal shows two lions supporting it and there is also a *Triratna* symbol between the lions worshipped by two devotees (male and female) who hold the garlands. The Brāhmi inscription of the Kuṣāṇa character is engraved on the pedestal which records that the image was installed in the year 23 of Kaniṣka's reign.

(4)   Next representative image of this age is an image of Bodhisattva, housed in the Lucknow Museum (No. 66.48). The inscription depicted on it, records that the image was made by *Śailarupakāra* i.e. the stone sculptor.[57] The image was discovered at Śrāvasti and furnishes the documental evidence to locate the Jetavana monastery. The first three lines of the inscription are incised in the early Brāhmi characters of the 1st century A.D. but the last line recording the Buddhist creed is a later addition of the 8th-9th century A.D. However, D.R. Sahni thought that the script is of pre-Kaniṣka period.[58]

(5)   A stele, housed in the Indian Museum, Calcutta (No. 25524), representing the Buddha figure in almost Katrā style Bodhisattva image, is a good representative image of the Kaniṣka period.[59]

---

53. *Ibid.*, fig. 80.
54. J. Ph. Vogel, *Catalogue of the Archaeological Museum at Mathura*, Allahabad, 1910, pp. 48ff.
55. *Ibid.*, p. 51.
56. R.C. Sharma, *Op. Cit.*, fig. 91, p.182.
57. *Ibid.*, p. 180, fig. 87.
58. D.R. Sahni, *Archaeological Survey of India Report.*, 1908-9, pp. 134,138.
59. R.C. Sharma, *Op. Cit..* fig. 85.

On the pedestal, however, the central lion is replaced by the *Dharmacakra* being worshipped by a couple and the other notable thing of this image is that no inscription is engraved on any part of the sculpture. The overall treatment of the image is rather weak and it seems that it is not a product of any skilled artist.

(6) A Headless Buddha image, housed in the Mathura Museum (No. 76.32), discovered at Govindnagar, is another representative image of this period. The treatment of the image shows the continuation of Yakṣa ideals i.e. bulky body, bulging breast etc.[60]

The characteristic features of the standing images of the Buddha/Bodhisattva of the Kaniṣka period.

The upper portion of the sculptures containing more or less similar features as that of the seated one. The standing images of the Buddha/Bodhisattva, bear the following features:

(1) The body of the sculptures are straight and stiff.

(2) The right hand is raised in *Abhayamudrā* and the left hand is held akimbo resting on the waist.

(3) The upper garment covers the left shoulder only with its fold. The lower garment reaches below the knee and its hem rests on the left hand. The notable thing of the lower garment is that it is fastened by a waist band which terminates into a double knot to the right side and it hangs down on the thigh.

(4) Generally a bunch of flowers is seen between the legs of the standing figures, sometimes it is replaced by a lion.

Some of the specimens of the standing Buddha/Bodhisattva images of the Kaniṣka period are as follows:

(1) A colossal image dated in the 3rd regnal year of Kaniṣka, housed in the Sārnāth Museum (No. B.1) is a good representative of the images of Kaniṣka period. The right arm of the sculpture is broken but the left hand is held akimbo on

---

60. *Ibid.*, fig. 82.

the waist. The sculpture was surmounted by a large parasol which has been recorded in the epigraph as *Chatrayaṣṭi*.[61]

(2)   Next representative image of this period is a life size image of the Buddha discovered near the Givindnagar mound. The image was brought to light by R.C. Sharma.[62] The image is housed in the Mathura Museum (No. 71.105). The head of the image seems to have been shaven, however, there is snail shell knot on it. The *Ūrṇā* bears five lines in relief. There is a *Cakra* on the palm of the right hand while finger tips show *Svastika* marks.

(3)   A railpost discovered at Jamalpur, housed in the Lucknow Museum (No. B.75), represents the *Karpardin* style of the Buddha image in the pose of *Abhaya mudrā*. The halo bears ornamental border. On the reverse side in the top panel a seated Buddha image is depicted in shaven head. The seat of the figure is supported by two lions.

(4)   Another railpost, housed in the Musee Guimet, Paris, depicts the similar type of Buddha figure as mentioned above. It is a tiny figure with the heavy impact of Yakṣa image.[63]

The characteristic features of the images of the Buddha of the Huviṣka period are as follows:[64]

The images of the Buddha from the Huviṣka period onwards show the penetration and culmination of the Gandhāran traits in Mathura school of art. This change might have caused by a strong influx from Gandhāra probably due to the fact that the art of this country by this time had risen to such a height that its product passed the border and drew the attention of sculptors from other parts of India.[65] Any way, let us survey the features:

(1)   The iconographic difference of Buddha and Bodhisattva figures emerges.

---

61. *Ibid.*, pp. 184-185, fig. 94.
62. *Ibid.* p. 183, fig. 92.
63. J.E. Lohuizen, *Op. Cit.*, fig. 30.
64. R.C. Sharma, *Op. Cit.*, pp. 188-189
65. J.E. Van Lohuizen, *Op. Cit.*, pp. 180-181.

(2)   Now the ear became elongated but the resemblance of the Yakṣa figure still continues.

(3)   The halo is more decorated and in addition to the earlier scalloped border it bears a beaded line or plain field is filled up by a full blown lotus.

(4)   The *Ūṣṇiṣa* on the head is more prominent and has more than one twist. It shows matted hair which is usually combed or show small curls.

(5)   The eyes of the sculptures became sharpened at the ends and a line on the neck is generally noticed.

(6)   The central lion on the pedestal is absent.

(7)   Additional thick plates are shown over the inner folds of the drapery.

(8)   The pedestal of the seated image invariably bears inscription and the central lion on the pedestal is disappeared; sometimes the pedestal bear a female figure in *Sāri*.

(9)   The traces of Gandhāran impact is noticed for the first time in the Mathura sculptures. The *Vajrapāṇi* of Gandhāran sculptures in the Scythian dress seems to be one of the attendants of the Buddha image.

Some of the specimens of the images of the Buddha which bear above-mentioned features are as follows:

(1)   A seated image of the Buddha discovered at Ahicchatra, housed in the National Museum, New Delhi (No. L.55.75) (Pl.1B.) is one of the best representatives of the images of the Buddha of the Huviṣka period.[66] The Buddha is shown in *Padmāsana* with the right hand raised in *Abhaya mudrā* and the left hand clenched on the knee. The Buddha is attended by *Vajrapāṇi* on the right and *Padmapāṇi* on the left side. The *Vajrapāṇi* is wearing a Scythian dress similar to *Vajrapāṇi* of Gandhāra images. The pedestal shows the worship of the *Bodhi*-tree by women who wore their *Sāris* in the Scythian fashion. The pedestal also bears

---

66.   D.L. Snellgrove (Ed.), *The Image of the Buddha*, New Delhi, 1978, p.57, fig.30.

a three lines inscription. It is probably the earliest dated image (110 A.D) which reflect the Gandhāran impact.

(2)   The next representative of this group is a seated image of the Buddha discovered at Ahicchatra, housed in the Boston Museum.[67] This image is very much similar to the above mentioned Ahicchatra image. The pleats of the drapery cover the left shoulder only. On the right side of the Buddha, *Vajrapāṇi* is depicted who has the thunderbolt in his right hand.

(3)   A headless seated image of the Buddha discovered at Govardhan housed in the Mathura Museum (No. 78.34) is the next representative of the images of this period. The Buddha is shown in *Padmāsana* and his right hand is turned in *Abhaya mudrā* while the broken left hand is clenched on the left knee. The sole of the feet shows various auspicious signs. The important stylistic changes are: additional thick pleats of the *Saṅghāti* appeared on the left shoulder, absence of lion or figurative expression on the pedestal, the frill of the lower garment are richly treated and the halo is large with a full blown lotus. The inscription is depicted on the pedestal which furnished probably the earliest archaeological document of the *Kāyasthas*.[68]

(4)   A Buddha head housed in the state Museum, Lucknow (No. J.226) is another representative of this group. Here the *Ūrṇā* between the eyebrows is prominent. There is multi-tiered top-knot on the head. The eyes are sharp at the ends. There is a marked line on the neck.

The standing images of the Huviṣka period do not show much difference from the earlier one, however there is stylistic overlap between Kaniṣka's phase and Huviṣka's early phase which are as follows:

(1)   The Yakṣa impact on the Buddhist images is now diluted and the images are comparatively less voluminous with a tendency of reducing the mass.

---

67.   J.E. Van Lohuizen, *Op. Cit.*, p. 172, fig. 32.
68.   R.C. Sharma, *Op. Cit.*, p.191.

(2)   The shoulders are brouder and less refined.

The specimen of this group:

(1)   A headless standing image of the Buddha discovered at Dhaulipiau near Mathura by R.C.Sharma, housed in the Mathura Museum(No. 80.1)is a representative of this group.[69] A bunch of lotus buds is seen over the detached hair knot which is placed between the legs. The lotus decoration is also seen behind the legs. The right arm is missimg and the left arm with thick pleats is resting on the waist and supports the hem of the scarf.

The characteristic features of the images of the Buddha belonging to the second phase of the Huviṣka's reign.

The images of the Buddha of the Mathura school of art witnessed another stage of development during the second phase of the Huviṣka's reign which indicates the Gandhāran impact on the Mathura school of sculptures.[70] The second phase of Gandhāran impact on the Mathura shows following new features:

(1)   The thick drapery covers both the shoulders and forming a 'V' shaped or semicircular pattern.

(2)   The feet of the deity are also covered with the garment and its broad pleats fall in a semicircular faishion.

(3)   The right hand is raised in *Abhaya mudrā* and the left hand holds the hem of the drapery which is almost parallel to the right hand.

Some of the specimens of the images of the Buddha which bear above mentioned features are as follows:

(1)   A standing image of the Buddha housed in the Mathura Museum (No. A-4) is one of the best representative of this period. The features of the image showing a shaven head, *Ūrṇā* between the eyebrows, there is scalloped carving on the edge of the halo. The right hand is raised in *Abhaya mudrā* and the left hand holds the hem of the drapery. The broad pleats are

---

69. *Ibid.*, fig. 103.
70. J.E.Van Lohuizen, *Op.Cit.*, p. 183.

seen in the drapery which covers both the shoulders and there is 'V' shaped formation of the *Saṅghāṭi*. below the necks which has two distinct incised lines.[71]

(2)   A headless Buddha image, housed in the State Museum, Lucknow(No. 14) is another representative of these group of images. The right hand of the Buddha is raised in *Abhaya mudrā* and the left hand lifted upto the armpit which supports the hem of the garment. The treatment of the drapery is similar to the above mentioned image.[72]

One of the important specimens of Buddhist sculptures of the Huviṣka period is a seated image of the Budddha discovered at Anyor, dated in the year 129 A.D., the image acquired its own style; this style influenced some of the sculptures. The style is known as Anyor idiom which form a distinct group.[73] The characteristics features of this group are as follows:

(1)   The drapery is coarse with thick pleats but despite heaviness and thickness transparent effect is retained to some extent.

(2)   The garment covers both the shoulders and forms a triangle.

(3)   The right hand is raised in *Abhaya mudrā* which is almost frontal.

(4)   The *Kapardin* style of hair is replaced by a row of notches or semicirular scratches.

(5)   *Dhyāni* postures appears for the first time.

(6)   The lions on the pedestal are usually frontal or inward.

Some of the specimens of the images of the Buddha which bear above mentioned features are as follows:

(1)   A seated image of the Buddha from Anyor, housed in the Mathura Museum(No. A.65), is one of the best examples of this group.[74] The right hand of the Buddha is turned in *Abhaya mudrā*; the position of palm is almost frontal instead of being in profile. The left hand is holding the hem of the *Saṅghāṭi*

---

71.   D.L. Snellgrove (Ed.), *Op.cit.*, fig. 32.

72.   R.C.Sharma, Op.Cit. , fig. 106.

73.   *Ibid.*, pp. 197-199; J.E.Van Lohuizen, *Op. Cit.*, pp. 188-189.

74.   R.C.Sharma,*Op.Cit.*, fig. 109.

which now shows somewhat thick and coarse pleats. The hair is arranged in the schematic notched fashion with rows of semicircular scratches placed one over the other. The lion carved on the pedestal is frontal instead of being in profile. The important addition on the pedestal is the depiction of a meditating Buddha as an object of worship between the two devotees. The drapery of meditating Buddha is the same as is worn by the presiding Buddha. On the lower rims of the pedestal, an inscription is engraved in Brāhmi characters which records the year 21.[75]

(2)   An image of the Buddha housed in the Boston Museum is a close resemblance with the Anyor Buddha image which might have been carved in the same period.[76]

The characteristic features of the images of the Buddha belonging to the last phase of Huviṣka's reign.

During the last phase of Huviṣka's reign, the Gandhāran impact became more prominent and the sculptures show some additional features which are as follows:[77],

(1)   The garment became more thick and stiff with broader folds and the 'V' shaped additional scarf go round the neck.

(2)   Besides *Abhaya* and *Dhyāna-mudrā*, other *mudrās* of the Buddha like, *Bhumisparśa* and *Dharmacakra* also emerged.

(3)   The hair became curly with top-knot and sometimes it is wavy like on Gandhāran Buddha images.

(4)   The halo became more elaborate besides scalloped border which shows shooting arrows.

(5)   The Bodhisattvas are frequently shown and now the flanking acolytes disappeared.

(6)   The events of the life of the Buddha became more prominent or popular.

---

75. *Ibid.*, p. 198.

76. J.E.Van Lohuizen, *Op.Cit.*, p. 197, fig. 40.

77. Ibid., pp. 201-201; R.C.Sharma, *Op.Cit.*, 200-203.

Some of the specimens of the images which bear above mentioned features are as follows:

(1)   A headless image of the Buddha discovered at Govindanagar and housed in the Mathura Museum(No. 76.19) is one of the best representatives of this period. The image is shown seated in *Dhyāna mudrā* which is carved in high relief. Neither animal nor inscription is engraved on the pedestal and the remaining field is left blank.

(2)   The next representative of this group is a seated image of the Buddha, discovered at Govindanagar and housed in the Mathura Museum(No. 76.17). The right hand of the Buddha is turned in *Abhaya mudrā* and the left hand holding the hem of the garment. The palm of the right hand is decorated with a double rimmed wheel. Here the fingers of the image is webbed (*Jālāṅgulikara*). The hair of the Buddha is made in schematic curl. There is *Ūrṇā* between the eyebrows. The almond shaped eyes are stretched upto the temples by a small horizontal stroke. The earlobes are rather small. The expression of face is sober. The halo is carved with a full blown lotus and decorated at the edge with shooting arrows etc. The treatment of the garment is similar as mentioned earlier.

The characteristic features of the images of the Buddha during the later Kuṣāṇa period.

During the later Kuṣāṇa period we find some stylistic change or difference in the images of the Buddha. Van Lohuizen [78] has noticed these changes on the basis of the two images of the Buddha, one from Śrāvasti, and the other from Sitalāghati which formed a separate group. The distinguishing features of the images of this group are as follows:

(1)   The 'V' shaped space formed by the garment below the neck now became roundish and cushion of *Kusa* grass is seen on the pedestal of the seated image.

(2)   The folded and stiff drapery on the right side is fan shaped.

---

78.   J.E. Van Lohuizen, *Op. Cit.*, p.201.

Some of the specimens of the images of the Buddha which bear above-mentioned features are as follows:

(1)   A small stone image of the Buddha discovered at Śrāvasti by John Marshall,[79] housed in the State Museum, Lucknow,(No. 66.183) bears the above mentioned features. The right hand of the Buddha is turned in *Abhaya mudrā*. The fingers of the Buddha are webbed (*Jālāṅgulikara*) and wheel is depicted on the palm. The halo is decoratead with a full blown lotus. The eyebrows are painted in such a way so as to create a cup like cavity for the eyes. On the pedestal there is a thick *Kusa*-grass cushion. A seated Bodhisattva in *Dhyāna mudrā* flanked by four worshippers who bear garlands in their hands, is depicted on the pedestal. On the lower rim of the pedestal an inscription is engraved which records that this is a pious gift of Sihadeva, a Prāvārika of Sāket.[80]

(2)   Another specimen of this group is a seated image of the Buddha discovered at Sitalāghāṭi (Mathura) and housed in the Mathura Museum [81](No.A.21). Here the Buddha is seated on a lion throne. The hands of the image are broken and the pedestal is partially damaged but the remaining portion shows that it has received almost the same treatment as mentioned in the case of the Śrāvasti Buddha image. Here the halo of the Bodhisattva on the pedestal, in comparison to Śrāvasti Bodhisattva image, is more elaborate.

The characteristic features of the images of the Buddha during the Huviṣka-Vāsudeva period:

The images of the Buddha in the last days of Huviṣka and early days of Vāsudeva created a stylistic crisis in the Mathura school. Consequently both Gandhāra and the indigenous ideals started dominating the images of this period. Here we notice the revival of the *Kapardin* form of the Buddha with *Uṣṇiṣa* and bare feet in *Padmāsana*. Ultimately the indigenous trend became more prominent

79.   J.Marshall, 'Mathura', *Archaeological survey of India*, Vol. X, 1910-11; R.C.Sharma, *Op. Cit.*, fig. 117.

80.   *ASR*, 1910-11, p.12.

81.   D.L.Snellgrove, *Op.Cit.*, p.56, fig.31.

and dominant and the Gandharan ideals start to decline. The alteration in drapery is more conspicuous which now became relaxed, less stiff, light and loose with independent folds though covering both the shoulders.

Some of the specimens which bear above mentioned features are as follows:

(1)   A headless small image of the Buddha discovered at Govindnagar housed in the Mathura Museum (NO. 76.33) wears the garment with loose and independent folds.

(2)   Another representative of this group is a headless image of the Buddha housed in the Mathura Museum (No.42.2919).

Here the seated Buddha is shown in *Dhyāna mudrā*. The garment shows relaxing trends and the folds became independent. The throne is supported by large lions and Bodhisattva who is being worshipped by the Śaka nobleman and woman. On the pedestal, there is a thick cushion of *Kusa*-grass. Thus, the pedestal bears many Gandhāran traits.

The characteristic features of the images of the Buddha during the Kuṣāṇo-Gupta phase.

The Kuṣāṇa rule came to an end after the death of Vāsudeva in 176 A.D.[82] which was followed by a period of political instability for more than a century until the Guptas came to power in 319-20 A.D.[83] The sculptures of Mathura school of this period have some distinctive features [84] which are as follows:

(1)   The massive body of the Buddha has thinned down and the facial expression shows some improvement which convey a feeling of serenity.

(2)   The thick and heavy folds in the drapery has given way to light treatment.

(3)   The treatment of the eyes become horizontal and earlobes become elongated.

---

82. H.C. Raychaudhury, *Political History of Ancient India*, Calcutta, 1972, p. 422.

83. R.C. Majumdar (Ed.), *The Vākāṭaka-Gupta Age*, Delhi, 1986, p.2.

84. R.C. Sharma, *Op. Cit.*, p.215.

(4)   The hair became curly and the halo become more decorated.

Some of the specimens which bear above mentioned features:

(1)   A headless seated image of the Buddha housed in the Mathura Museum (No. 13.361) is one of the best representatives of this group. Here the Buddha is shown in *Dhyāna mudrā* and seated on a lion throne. The drapery covers both the shoulders which shows the Kuṣāṇa trend but it is considerably relaxed and its independent folds make it transparent. The neck bears three round lines *(Trivalī).*[85]

(2)   The next representative of this group is a seated image of the Buddha discovered at Bodhgayā, housed in the Indian Museum Calcutta (Acc. No. A25023).[86] The transparent garment covers half shoulder only. The image shows some Guptan ideals viz., the neck bears three lines round the neck, curly hairs, elongated earlobes, enlarged half closed meditating eyes, serene facial expression and improvement in the modelling of the body. The inscription on the pedestal records that it was consecrated in the year 64 of some Mahārājā Trikamaladeva.

The characteristic features of the images of the Buddha during the Gupta period.

Aesthetically the images of the Buddha reached its zenith during the Gupta period. The images of this period became slim and slender and the expression reflected the inner bliss and serenity. But the production of sculptures became slowed down, may be due to the decline of trade and commerce and the lack of royal patronage, because the Guptas ruled from Pāṭaliputra. However, Mathura enjoyed the patronage of traders and teachers. The sculptors of this period concentrated their mind in the quality rather quantity.[87] The images of the Buddha of this period bear the following features:[88]

---

85.   *Ibid.*, fig. 132.

86.   F.M. Asher, *The Art of Eastern India* (300-800 A.D.) Delhi, 1980, p. 19, pl. 11; D.L. Snellgrove (Ed.), *Op. Cit.,* fig.28.

87.   R.C. Sharma, *Op. Cit.*, p.140.

88.   V.S. Agrawala, *Gupta Art*, Varanasi, 1977, p.25.

(1)   Emphasis has been given on the expression. The physical forms and the facial expressions became harmonised and they express the feeling of serenity, inner bliss, calmness, compassion, and the other sublime feelings.

(2)   The images which fall in the period upto fourth century or early fifth century still bear some Kuṣāṇa ideals but by the fifth century A.D. they became gradually more slender and elegant and graceful. The limbs are longer, delicate, flexible and less muscular. The torso and limbs harmonize with the facial expression.

(3)   The face is generally oval and the head is covered with small spiral locks of hair including the *Uṣṇiṣa*. The ears are elongated and eyes are half closed with gaze. The eyebrows are long and arched and in the case of the images made by Yasadinna, they meet each other. The eyes are made large and horizontal, lotus bud-shaped, half open, infusing inward vision and meditation. The lips are full and clearly marked upto the end of the mouth with their upper lip and drooping to lowerlip. The double round chin shows inner strength.

(4)   Now the drapery becomes light, diaphanous with wet-silk like transparency which shows the outline of the body. The garment generally covering both the shoulders fall in schematic loops or ridges down the central vertical line shows the natural body. The ends of the upper garment are held in the left, falls in frills.

(5)   The animals and vegetable world finds no importance and even the human beings have a minor role in the composition and the Buddha figure dominate all others.

(6)   The beautiful double marking on the neck makes it *Kambugriva* (conch shell like neck).

(7)   The aim of installing the image or images is often recorded as *Anuttarajñānavāpti* (attainment of supreme knowledge) as against the '*Sarvasttvānāmhitasukhārtham*' as mentioned in the Kuṣāṇa image inscriptions. This suggests a march from mortality to divinity or towards the Mahāyāna ideal.

The images of the Buddha during the Gupta period reflect gradual slow process of change which can be divided into three phases:

1.　　Phase one, covering the period of early Gupta reign.

2.　　Phase two, covering the period of Kumāragupta - Skandagupta.

3.　　Phase three, covering the period of the last days of the Gupta reign.

PHASE - 1 : The images of the Buddha of this phase still retain some earlier features such as squat body, flowers between their feet, raised right arm, wheel on the palm, shelved drapery on the left arm, placement of the feet together etc. Of course, there is slight change in the halo which besides 'Solar disc' of early period also includes several garland band of late Kuṣāṇa period with a ring of leaves.[89] The halo became common in later Gupta period. The Gupta features are also noticed in the treatment of half closed eyes. The inward gaze of the images change its status and impression. This change might have taken place in the last decades of the fourth century A.D.[90]

Some of the specimens of the images which bear above mentioned features are as follows:

(1)　　A seated image of the Buddha housed in the Indian Museum Calcutta (Acc. No. A25023)[91] combines the Kuṣāṇa idiom with inner spiritualism of the Gupta period.[92] The inscription on it of the year 64 refers to some Mahārājā Trikamaladeva. S.K. Saraswati accepts this image to be of Mathura origin,[93] while Harle dates it to 384 A.D. and of eastern Madhyadeśa style.[94] Due to it being made of a yellowish buff stone, William considers it a local product of Bihar.[95]

89.　J.G. Williams, *The Art of Gupta India - Empire and Provinces*, Delhi, 1983, p. 30.

90.　*Ibid*.

91.　D.L. Snellgrove, *Op. Cit.*, fig. 28.

92.　S.K. Saraswati, *A Survey of Indian Sculptures*, Calcutta, 1957, p. 133.

93.　*Ibid.*, p.134.

94.　J.C. Harle, *Gupta Sculpture*, Oxford, 1974, p. 33.

95.　J.G. Williams, *Op. Cit.*, p.33.

(2)   The next representative of this group is a standing image of the Buddha, discovered at Govindnagar, Mathura. It has been dated to about 390-400 A.D. The image wears foldless drapery which we generally find in images of the Sārnāth school of art. The rest features of the image is the same as discussed above.

PHASE - 2 : In this phase the Kuṣāṇa features on the images of the Buddha dissappeared and the body and head of the images became more elongated. The garment consists entirely of string folds, which are arranged symmetrically on the torso to produce a varied yet harmonious pattern. The feet are set farther apart, their arch higher and no object occurs between them. The right leg of the standing figures are very slightly bent giving the figures a relaxed look. The features of the images are sharp and clearly engraved and full of sprituality. The rays in the halo are replaced by a large lotus, a band of leaves, beads and other vegetable motifs. However, the seated images of the Buddha are not very much common with the standing ones.[96]

Some of the specimens which bear above mentioned features are as follows:

(1)   A standing image of the Buddha discovered at Jamalpur and housed in the Mathura Museum (No.A.5) is a remarkable specimen of this group.[97] The light and transparent garment looks like a thin muslin. The hair is arranged in fine curls, the shape of the eyes are like half open lotus buds and the ears are elongated. The halo is richly decorated with several ornamental bands beginning with the full blown lotus at centre, successively encircled by a wreath issuing from crocodile heads, row of rosettes, and extemely beautiful band showing stylised peacocks or gees intervened by full blown lotuses, a twisted wreath with a beaded line and lastly the scalloped edge. Unfortunatley half portion of the right hand and the tip of the nose from the right side is broken. The forearm of the right hand is broken, however the remaining portion indicates that

---

96. *Ibid.*, p.70.
97. D.L. Snellgrove, *Op. Cit.*, p.94, pl.56.

it was turned in *Abhaya mudrā*. Above all, the expression suggests a combination of serenity and divine bliss. There are two lines of inscription engraved on the pedestal which records, *inter alia*, the name of sculptor-donor, Yasadinna.

(2)   A similar image from the same place is housed in the Rāṣṭrapati House (Ashoka Hall, pl.3).[98] Here the nose of the image is intact and there are some variations in the carving of the halo which begins from the knob of a full blown lotus, floral bands, undulating rich scroll, twisted wreath beaded line, and lastly, the scalloped edge. The difference is that the sylised peacocks or geese motif is replaced by the scroll. Another difference is that here an inscription is depicted on the pedestal.

(3)   A Buddha head discovered at Chamuṇḍā mound and housed in the Mathura Museum (No. 49.3510) is the next representative of this phase.[99] Aesthetically, it is a superb piece with fine curly hair, elongated earlobes, lotus shaped half open eyes with long eyebrows which meet with the nose to form three perfectly equal angles. The lips are full and the nose is thin. The overall facial expression imparts the feeling of serenity and contemplation. This is the creation of skilled sculptor Dinna.

(4)   The next representative of this phase is a Buddha image discovered at Jamalpur and housed in the Indian Museum Calcutta (No. 254-61).[100] The image is similar to the Jamalpur Buddha image as described above, however, the features of this image are less impressive.

One of the other important images of this phase are the images of Bodhisattva which reflect different ideals.

PHASE - 3: During this phase, the Mathura school of art produced a few good images of the Buddha. The invasion of the Huṇas proved to be a big blow to the Mathura school of art. The invasion led to the break-up of communication of Mathura with

---

98.   N.Ray, *Idea and Image in Indian Art*, Delhi, 1973, fig.15, R.C. Sharma, *Op. Cit.*, pp.222-223, fig. 141.

99.   D.L. Snellgrove (Ed.), *Op. cit.*, p.94, fig.54.

100. *Ibid.*, fig.57; N.G. Majumdar, *A Guide to the sculptures in the Indian Museum*, part II, Delhi, 1937, p.82.

neighbouring areas. Simultaneously, Śārnāth emerged as an important centre of art and the break up of the Gupta Empire contributed to the fall of the Mathura school of art. Actually the fall of Mathura school of art had started after the decline of the Kuṣāṇa empire as mentioned earlier, that during the Gupta period Mathura school of art had produced quantitatively less images of the Buddha.

The characteristic features of the images of the Buddha belonging to phase - 3 are as follows:

(1)   Images of this period are less impressive.

(2)   The Yakṣa type ideals or models revived.

(3)   Apart from the images of the Buddha, Mathura school of art of this period also produced the images of the *Dhyāni* Buddhas and different varieties of Bodhisattva [101] etc., though this type of tendency had already began with the Kuṣāṇa rule.

Specimen of the image of this period which bears above mentioned features:

(1)   A standing image of the Buddha discovered at Mathura and housed in the State Museum, Lucknow, [102]shows that it is influenced by the traditional Yakṣa images. Aesthetically it is less impressive than the images from Jamalpur. It may be a poor imitation of an earlier model.

---

101. Gregory Schopen, "The Inscription on the Kuṣāṇa Image of Amitabh and the character of the Early Mahāyāna in India," *JIABS*, Vol. 10, No.2, 1987, pp. 99-101.
102. D.L. Snellgrove, *Op. Cit.*, p.94, fig. 58.

CHAPTER - 3

ICONOGRAPHY OF THE BUDDHA IMAGES IN GANDHĀRA SCHOOL OF ART

3.1 *GANDHĀRA : Historical perspective of the place.*

3.2 *Buddhism in Gandhāra*

3.3 *Gandhāra school of Buddhist art*

3.4 *Iconography of the Buddha images*

3.1 *GANDHĀRA : Historical perspective of the place:*

Gandhāra was situated in the North-western frontiers of undivided India. Now it is the part of Pakistan. It denotes the region comprising the modern district of Peshawar and Rawalpindi etc.

In the *Rigveda* Gandhāra is mentioned as a producer of fine quality of wool.[1] The *Aṅguttaranikāya* mentioned it as one of the sixteen *Mahājanapadas*.[2] The ancient name of this place, Gandhāra or Gandhāri is also mentioned in the *Aitareya* and *Śatapath Brāhmaṇ*, the *Mahābhārata* and the *Jātakas*.[3] Apart from these, Gandhāra is also mentioned in some foreign accounts.[4] The river Indus divided the whole Gandhāra region into two parts: eastern and western.[5] Taxila (Rawalpindi Distt. of Pakistan) was the main city of eastern Gandhāra and Pushkalāvati (Peshawar Distt. of Pakistan) was the chief city of western Gandhāra. The entire Gandhāran area was divided into seven parts, Taxila, Pushkalāvatī (Chārsadda), Nagarahāra, Swāt valley

---

1. *The Rigveda*, I.26.7.

2. *The Aṅguttaranikāya*,1.i.213.

3. J. Goswamy, *Cultural Historical of Ancient India*, Delhi, 1978, p.4; B.N. Chaudhury, *Buddhist Centres in Ancient India*, Calcutta, 1982, p. 137.

4. J.Goswamy, *Op. Cit.*, pp.4-5.

5. H.C. Raychaudhury, *Political History of Ancient India*, Calcutta, 1950, p.59.

(Udyāna), Kāpiśa, Bāmian and Bālkh etc.[6] The Gandhāra school of art largely flourished in the border land of modern Pakistan and Afghanistan, bound on the east by Taxila, on the west by Kābul and Begram (ancient Kāpiśa), on the north by Kashmir and ancient Udyāna, and on the south by Bannu and Waziristan. Later on, it spread further to Sindh, and Baluchistam and the ancient silk-route from Kunduz on the Oxus in North Afghanistan as far North-east as Mirān in the Tarim basin of Central Asia.[7]

Geographically Gandhāran region was so situated that it lay exposed to all sorts of foreign contacts and influence - Persian, Greek, Roman, Śaka and Kuṣāṇa etc. Subsequently it gave birth to a syncretic cosmopolitan culture that found manifestation in the Gandhāra school of art.

In the 6th-5th century B.C. Gandhāra was the part of Achaemenid empire of Persia and for a short period in the 4th century B.C. it was occupied by the armies of Alexandar. Later on, it was conquered by Chandragupta Maurya. But after the disintegration of the Mauryan empire, Gandhāran region was occupied by the Indo-Greeks, the Śakas, the Scythians and the Kuṣāṇas, respectively. The age of the Kuṣāṇas proved to be a glorious one for Gandhāra when its art reached its zenith. In the third century A.D. Gandhāra again came under the Persian rule[8] and in the fourth century A.D., Gandhāra was re-occupied by the later Kuṣāṇas. In about 465 A.D., the white Huṇas conquered the Gandhāran region who are credited to destroy the Buddhist establishments in that area.[9] Sun-Yung a Chinese pilgrim who visited Gandhāra in 520 A.D., when the Huṇas were ruling there, vividly mentioned about the destruction of the Buddhist monasteries, *Stūpas* etc. by the Huṇas.[10] The Huṇas were supplanted

---

6.   B.B. Shrivastava, *Prāchina Bhārtiya Pratimā Vijñāna evam Mūrtikalā* (Hindi), Varanasi, 1981, p.336.

7.   R.C. Majumdar, *Ancient India*, Delhi, 1974, p.123; K.Deva, "In Defence of Gandhāra Art", in M.S. Nagaraju (Ed.), *C. Sivaramamurti Commemoration Volume*, Vol.I, Delhi, 1987.

8.   R.C. Majumdar (ed.), *Op. cit.*, p. 123.

9.   *Ibid.*

10. R.C. Majumdar (ed.), *The Classical Age*, Bombay, 1954, pp. 34-35; S. Sengupta, *Buddhism in the Classical Age*, Delhi, 1985, pp. 71-72; F. Sehrai, *The Buddha story in Peshawar Museum*, Peshawar, 1978,p.4.

by the Hindūshahiyyā (or Hindushāhi) kings of Huṇḍ who ruled over the Gandhāran region upto the 11th century A.D. when they were defeated by Md. Ghazni.

Thus, due to the peculiar geographical position, Gandhāra became the meeting place of differnt races of the world including India which resulted in the birth of a cosmopolitan culture. The language of the Gandhāran people were most probably the North-western variety of Prākrit and they used Kharoṣṭhi and sometime Brāhmi script also for their writing.[11] Apart from indigenous language they had also adopted some foreign languages including Greek, which was popular, as attested by the languages used in various coins and inscriptions discovered in the Gandhāran region.[12]

The fusion of these divergent races are reflected in the religion and faith of the Gandhāran people. The coins of Kuṣāṇa kings, Kaniṣka and Huviṣka who had their capital at Peshawar, shows a galaxy of gods and goddesses which also include the Iranian deities like, the Sun-god (*Miiro* or *Mioro* (Persian, *Mithra, Mihr*) (Indian, *Mitra, Mihira*), the Moon-god(*Mao*), the Wind- god (*Oado*) (Persian, *Vado*, Indian, *Vata*)), the Fire-god(*Athsho (Persian, Atash)*), the War-god(*Orlagno*(Persian, *Bahram*)), Victory-god *Oanindo* etc.[13] Greek deities like *Heracles, Helios* and *Selene* etc. Indian deities were *Śiva(Oesho), Umā, Skanda-Karttikeya, Visākha, Mahāsena*, and Buddha etc.[14]

The political expansion of the Kuṣāṇas, particularly under the first three kings, was remarkable one. The Kuṣāṇa territory reached its culminating point under Kaniṣka. The vast area comprising of Afghanistan and adjoining parts of Central Asia, Pakistan, Northern India upto Pāṭaliputra came under their control.[15]

---

11. J.Marshall, *The Buddhist Art of Gandhara*, Cambridge, 1960, pp. 1-2; *Bulletin of the School of Oriental and African Studies*, Vol. XXIII, p.47.

12. Bhaskar Chattopadhyaya, *The Kusāna State and Indian Society*, Calcutta, 1975, pp. XXIX, 228-248.

13. R.C.Majumdar(ed.)*The Age of Imperial Unity*, Bombay, 1953, p. 147,150.

14. *Ibid*.

15. Bhasker Chattopadhayaya, *Op.Cit, p. XXIV.*

The Kuṣāṇa period witnessed a rapid growth in trade and commerce. Both internal and external trade routes either passed through the Gandhāran region or it was linked with the region. The *Jātakas* and the accounts of Megāsthenese mention that the main trunk road that covered "the entire span of Eastern and Northern India from the port of Tāmralipti to the frontier town of Pushkalāvati, going by way of Campā, Vārānasi, Kausambi, Mathura, Sāgala, and Takṣaśilā."[16] From Pushkalāvati, the trade route proceeded further through the Khyber Pass, the Kabul Valley, Haḍḍā, Nagarahāra, Bāmiyan and Hindukush and terminated at Bactria.[17] Bactria was a sort of cross road of the routes from India, China through Central Asia and Mediterranean world. Towards the west from Bactria, the routes proceeded " through the northern part the Persian Desert, the grassy downs of area (herat) and the Zagros valley etc. across the Tigris and the Euphrates to Syria. Besides, the existence of the two other routes from Bactria to the west was also cited.[18]

The trade contact between India and the western countries was brisk during the Kuṣāṇa period which may have contributed to the intermingling of religious faith and forms of diverse races, a process which was stimulated by the numismatic tradition built up by the Kuṣāṇa.[19] Gandhāra being the important centre of the Kuṣāṇa empire it benifitted most from the growing trade and commerce which resulted in prosperity. The Kuṣāṇa rulers, despite having their own faith, maintained tolerant and impartial attitude towards all existing religions.[20] The prosperity of Gandhāra was due to the above political and economic factors from Indo-Greeks to the Kuṣāṇa period that resulted in the development of religions, art and architecture.[21]

3.2   Buddhism in Gandhāra:

It is very difficult to determine when exactly Buddhism was spread in the Gandhāran region for the first time. The early Buddhist

16. *Jātaka*,III, 365; G.L.Adhya, *Early Indian Economics*, Bombay, 1960, p.103.
17. G.L.Adhya, *Op.Cit.*, p. 155.
18. *Ibid.*, pp. 104-105.
19. Bhaskar Chattopadhyaya, *Op.Cit.*, p. 175.
20. *Ibid.*, p. 176.
21. *Ibid.*, pp. 176-178.

texts are quite silent about the early spread of Buddhism in this region, though they record frequent visits of scholars and merchants from central India to this region. Taxila, then capital of Gandhāra, was a famous centre of learning in various sciences and arts. The *Papañca Sudani*[22] mentions that Bimbisāra of Magadha had friendly relation with the king of Gandhāra, Pukkusāti and once Bimbisāra sent him a letter containig the news of the appearance of the Buddha, the Dhamma, and the Saṅgha in the world. Getting this news, the king became so impressed that he decided to be a convert in Buddhism and in a short period he ordained himself as a monk. Subsequently he left his kingdom and went to Śrāvasti to see the Master. But except this story we do not have any further information about the traces of Buddism in this region uptill the appearance of Majjhantika in the third century B.C.

The *Mahāvaṁsa*[23] mentions that after the completion of third Buddhist council, held at Pāṭaliputra, Buddhist missionaries were sent to different parts of the country to propagate the Dhamma. *Thera Majjhantika* was the incharge of the Kashmir-Gandhāra region for this mission. At the time of his visit, the people of Gandhāra were being harrased by the Nāga-King *Aravāla*. Majjhantika by his miraculous powers subdued the Nāgas. Subsequently the Nāga-king together with his followers, the *Yakkha Paṇḍaka* and his wife *Hāriti* were converted into Buddhism.[24] A similar tradition is also found in the Tibetan *Dulvā*(Vinaya), *Rājatarangini*, *Bu-Ston*, *Aśokāvadāna* etc.[25] The *Rājatarangini* records the existence of two pre-Aśokan monasteries at Jalora and Saurasa built by king Surendra.[26] But this record has yet to be proved by the archaeological evidence. It is more probable to say that Buddhism was widely spread all over the Kashmir-Gandhāra region during the reign of Aśoka. Both Kalhaṇa and Huien-Tsang mentions about a number of *Stūpas* built by Aśoka in the North-western region.[27] However, Huien- Tsang mentions that he had

---

22. *DPPN*. Vol. I, Delhi, 1983, pp. 748-749.

23. *Mahāvaṁsa*, XII, p. 9ff, edited by Geiger (PTS).

24. *Ibid*.

25. Quoted from A.C. Banerjee, *Sarvāstivāda Literature*, Calcutta, 1957, p.91.

26. *Rājatarangini*, I verses 94, 118 (Ed.), M.A.Stein, 2 Vol., Varanasi, 1961.

27. *Ibid.*, 101-102; S.Beal, *The Buddhist records of the Western World*, London, 1906, p. XXXii.

seen only a few *Stūpas* erected by Aśoka near Pushkalāvati. But there is no definite information regarding the shape of the Aśokan *Stūpa* at Taxila, the only available indication of its site being that of a vast Buddhist establishment, with an enormous *Stūpa* as the nucleus has been called *Dharmarājikā* in inscriptions found at the site. Shāhbāzagarhi and Mānsehrā bear the Kharoṣṭhi versions of the fourteen Rock Edict of Aśoka on which rules of conduct are engraved and which he deemed to be most conducive to the welfare of his people.[28] The *Stūpa*-cult spread by Aśoka had a great impact on these *Stūpas*.[29]

After the disintegration of the Mauryan empire, Gandhāra passed into the hands of Indo-Greeks who embraced Buddhism cordially. The most famous among the Indo-Greek rulers was Menānder(Pāli-Milinda) who favoured Buddhism and probably got converted also.[30] Milinda or Menānder is immortalized in one of the main characters of the text, the *Milinda Pañho* and the text also suggests that he got converted into Buddhism.

He had his capital at Sāgala (modern Sialkot). Besides, the usage of the title *Dikaiou* on the obverse with its Prākrit equivalent *Dharmikas* on the reverse of some of a few silver and copper coins, and the adaptation of the "wheel" symbol also appear to show his Buddhist leanings, though it may not necessarily mean his conversin to the faith, considering the occurence of the same title in the coins of some other Indo-Greek Kings.[31] An inscription on a relic casket discovered at Shinkot in Bajaur region (Swāt) records that in the reign of Menānder, Viyākamitra[32] enshrined the corporeal remains of Sākyamuni Buddha endowed with life (*Prāṇa-sameda-śarira*) in a casket, and Vijayamitra, who may have been the son or grandson of Viyākamitra, re-established the sacred remains and made arrangements for their regular worship.[33]

---

28. D. Mitra, *Buddhist Monuments*, Calcutta, 1971, pp. 10-115.
29. J.Marshall, *Op.Cit.*, p. 3.
30. A.K.Narain, *The Indo-Greeks*, Delhi, 1980(reprint) pp. 97-99.
31. A.N.Lahiri, *Corpus of Indo-Greek Coins*, Calcutta, 1965, p. 160.
32. N.G.Majumdar considers him as a subordinate king under Menander, N.G.Majumdar,*EI*, Vol.XXIV, 1937, p. 1ff.
33. R.C.Majumdar(Ed.), *Op. Cit.*, pp. 114-115.

Apart from the Greek rulers, Greek people also patronised Buddhism and some of them got converted to Buddhism. A Greek named *Theodoros* placed a casket containing the sacred relics of the Buddha for the good of a large number of people in the Swāt valley.[34]

After the Indo-Greeks, Gandhāra came under the suzerainty of the Śakas and the Scythians respectively who followed the policy of Indo-Greeks in patronizing Buddhism. The Taxila Copper-plate inscription dated in the reign of Śaka king Maues(Ist century B.C.) records the enshrinement of the relics of the Buddha and the erection of a monastery by certain Patika, son of a Śaka Kṣatrapa.[35] The Kalwan Copper-plate inscription of the year 134 and Taxila silver scroll inscription of the year 136 suggests that Buddhism was patronised and practiced by the people at the time of Azes II.[36] By this time Sarvāstivāda school of Buddhism had already established their strong holds in the North-western region and also was patronized by the Śakas.[37] The coin of Śaka king Maues depicts a cross-legged human figure which has been interpreted by some scholars as the earliest depiction of the Buddha.[38] (It has already been discussed in the very first chapter).

After the fall of the Śakas and the Scythians, Gandhāra came under the control of the Kuṣāṇas who were great patrons of Buddhism who helped in the spread of Buddhism to Central Asia and even further. The depiction of seated human figure on the coins of Kujula Kadaphasis (15-65 A.D.) with Kharoṣṭhi legend on it has been interpreted by some as the image of the Buddha.[39] Kujula was succeeded by Vima Kadaphises whose coins contain the figures of Śiva and trident etc. So we may surmise that he was influenced by the Śaiva-cult. Kaniṣka, whose place in the history of Buddhism is well known, came to the throne after Vima Kadaphises.[40] He (Kaniṣka)

34. K.L.Hazra, *Royal Patronage of Buddhism in Ancient India*, New Delhi, 1984, pp. 118-121.

35. *Ibid.*, pp. 122-123.

36. *Ibid.*, p. 123.

37. *Ibid.*, p. 128.

38. A.K.Narain, "First Images of the Buddha and Bodhisattva..." in *Studies in Buddhist Art of South Asia*, Delhi, 1985, pp. 1-21.

39. K.L.Hazra.*Op. Cit.*, pp. 134-135.

40. *Ibid.*, pp. 135-155.

constructed a huge *Stūpa* and a monastery at Peshāwar which has been described in various accounts and literature.[41] Kaniṣka was instrumental in convening the fourth Buddhist council which was held at Kashmir or Jalandhar under the presidentship of Vasumitra in which extensive commentaries were composed on the *Sūtra, Vinaya,* and *Abhidharma* in Sanskrit. Kaniṣka's coins (Gold and Copper) containing the figures of the Buddha and Maitreya with the Greek legend 'BODDO', SAKAMANO BOUDO, METRAGO BOUD. Maitreya is recognised as the Buddha and not Bodhisattva.[42] Prolific figures of the Buddha and Bodhisattva were made during his period both at Gandhāra and Mathura. As discussed earlier, during the Kaniṣka period Buddhism spread into Central Asia and China. But this credit can not be given to Kaniṣka alone, because Indian merchants, who used to travel eastern Turkestan and among whom most of the merchants were Buddhists, were not less responsible for the propagation of the message of the Master.[43] Of course the king provided facilities on the trade routes such as security arrangements etc. The economic prosperity of the period resulted in the erection of many Buddhist monasteries, *Stūpas*, images etc. The successors of Kaniṣka I, Vāsiṣka (102-106 A.D.) and Huviṣka (106- 108 A.D.) etc. were also the patrons of Buddhism. Vāsiṣka during his short tenure built a monastery at Janakpura.[44] Huviṣka had constructed a monastery at Hushkapura in Kashmir. The Wārdak Vase inscription (Afghanistan) of the year 51 records that the relics of the Buddha were established at Vagramarega Vihār in a *Stūpa*.[45] Another emperor Kaniṣka II, who also ruled for a short period patronised Buddhism. Vāsudeva I (145-176 A.D.), though a Śaiva, adopted tolerant policy towards Buddhism.[46] Thus, the whole Kuṣāṇa period may be considered as important in the history of Buddhism.

---

41. K.W.Dobbins, *"The Stūpa and Vihāra of Kaniṣka I*, Calcutta 1971, pp. 12-23,33-44.
42. K.D.Bajpai,"Buddha and Maitreya in early Indian Art", in Devendra Handa (Ed.), *Recent Studies in Indology*, Vol. II, Delhi, 1989, pp. 348-349.
43. P.V. Bapat (Ed.), *2500 years of Buddhism*, Delhi, 1976, (reprint), pp. 58-59.
44. K.L. Hazra, *Op. Cit.*, pp. 155-156.
45. D.C. Sircar, *Select Inscriptions*, Vol. I, Calcutta, 1942, pp. 159-160.
46. K.L. Hazra, *Op. Cit.*, pp. 159-160.

The condition of Buddhism after the fall of the Kuṣāṇa power is not very much clear, mainly due to the paucity of concrete materials. It seems that the rulers who governed over this region as after the Kuṣāṇas, did not show much inclination towards Buddhism.[47] However, Buddhism did not totally extinguish from the Gandhāran region at least upto the 7th century A.D. which is proved by the records of the Chinese travellers as well as by the findings of archaeology. Fa-Hien found the *Dharmarājikā vihār* at Taxila, which was established by Kaniṣka, in a flourishing condition in the 4th-5th century A.D. He has referred a monastery at Purushpura (modern Peshāwar) with more than 700 monks.[48] In the Haḍḍā (in the Jalalabad district) he saw a *Vihāra* where "the flat bone of Buddha's skull is said to have been deposited which is regularly worshipped by the king of the country."[49] Another Chinese traveller, Sung- yung who came to India in early 6th century A.D., and visited only the North-west borderlands of India, mentioned about another holy site situated to the South-east of Purushpura where he saw a sacred *Pipal* or *Bodhi* tree which was planted by Kaniṣka over the spot where he buried a copper vassal.[50] Fa-Hien has escaped to mention it. Huien-Tsang reports to have seen numerous *Stūpas* and monasteries in the Gandhāran region, though they were in somewhat decaying condition.[51]

Archaeological finds suggest that *Stūpas* and other buildings continued to be erected around the Dharmarājikā *Stūpa* from the 4th to 7th century A.D. The Bhamala monastery in Taxila owes its origin to the period of 4th or 5th century A.D.[52] At the end of 5th century A.D. as mentioned earlier the Huṇas occupied Gandhāra and destroyed the Buddhist establishments.

Now the question arises what form of Buddhism was prevalent in Gandhāra through the ages? As it is obvious that the faith

---

47. S. Sengupta, *Buddhism in the Classical Age*, Delhi, 1985, p. 16.
48. S. Beal, *Buddhist Records of the Western World*, London, 1906, p.xxxiii.
49. J. Legge, *The Travels of Fa- Hien...*, Oxford, 1886, p.36.
50. S. Beal, *Op. Cit.*, p. CV.
51. *Ibid.*, p. xxxii.
52. S. Sengupta, *Op. Cit.*, pp.16-17; J. Marshall, *Taxila*, Vol.I, Cambridge, 1951, p. 119.

propagated by Majjhantika in Kashmir-Gandhāra area was Theravāda. Even during the reign of Kaniṣka, Hīnayāna was dominant in this region. The fourth Buddhist council, held at Kashmir or Jalandhar was dominated by the Sarvāstivādins, a school of the Hīnayāna. There is hardly any evidence of Mahāyāna Buddhism being patronised by Kaniṣka. Both Fa-Hien and Huien-Tsang mention that Hīnayāna was the predominant school of Buddhism in Gandhāra and Kabul valley.[53] This is further corroborated by the inscriptions of this region. The inscription of the year I of Kaniṣka on the reliquary discovered from Shāh-ji-ki-dheri,[54] the Zedā inscription of year 11 (139 A.D.) near Uṇḍ[55] and the Kurram Casket inscription of the year 20 (148 A.D.)[56], relates the gifts etc. to Sarvāstivādin teachers. Some inscriptions belonging to this region like the Bedadi copper ladle inscription (168 A.D.),[57] the Taxila copper ladle inscription,[58] and Palatu Dheri Jars inscription[59] also refer to the gifts to the Hīnayāna sect. But Huien-Tsang mentions that Taxila was the centre of Mahāyāna Buddhism.[60] Thus, we may surmise that both Hīnayāna and Mahāyāna tradition of Buddhism flourished in the Gandhāran region.

3.3   Gandhāra school of Buddhist art:

The Gandhāra school of art has a distinct place in early Indian art history. The art findings of this school are mainly Buddhist in origin. The sculptures discovered at Gandhāra are both in round and in relief. The sculptures in relief mainly relate the events of the life of the Buddha, the *Jātaka* stories etc. These Buddhist themes were depicted around and on small and medium sized *Stūpas* in monasteries, with imaginative bas-reliefs showing the Buddha in human form. Later on, the Buddha was represented in high relief along with the base

---

53.   Y. Krishan, "Was Gandhara Art a Product of Mahāyāna Buddhism? in *JRASGBI*, 1962, pt. 3-4, pp. 106-108.

54.   Sten Konow, Kharosṭhi Inscriptions,Vol.II, Varanasi, 1969, p. 137. pl.XXV.

55.   *Ibid.*, pp.142-145, pl. XXV.2

56.   *Ibid.*, pp.152-155, pl. XXVII, XXIV.

57.   *Ibid.*, pp.88-89, pl.XVII.4

58.   *Ibid.*, pp.87-88, pl.XVV.3

59.   *Ibid.*, pp.120-122.

60.   Quoted from S.Sengupta, *Op. Cit.*, Chapter-3, p.73.

of the *Stūpas* or steles to be kept in private chapels and statues in monasteries.

The first specimen of this school of art was discovered by Genard in 1832 A.D.[61]

NOMENCLATURE : The Gandhāra school of art has been called by various names viz., Indo-Bactrian, Indo-Greek or Graeco-Buddhist and Indo-Roman or Roman-Buddhist etc., mainly by the western scholars. Dr. Gottlieb Welhem Leitner (1840-99) for the first-time coined the phrase 'Graeco-Buddhist' for the sculptures of Gandhāra[62] because its technique and style seemed to be influenced by Hellenistic art and its theme was Buddhist.[63] Terms like Indo-Hellenistic, Graeco-Roman has also been applied for Gandhāra school of art. The inspiration and theme of this art form was Buddhist and various Buddhist symbols have been observed as purely Indian element in Gandhāra art. The representation of Mithraic cult from Persia and animal motifs like stepped merlon, winged lion, campaniform capital with or without addorsed bulls and animals with human heads in Gandhāra art is said to be an Iranian influence.

The sculptures in relief which illustrate the various incidents of the life of the Buddha are very much in chronological sequence, in comparison to the reliefs of Bharhut, Sānchi, Amarāvati and Nāgārjunakoṇḍa etc.[64] The representation of Maya's Dream, for example, is usually followed by the interpretation of the Dream, the Birth in the Lumbini grove, the return to Kapilavastu, the reading of the Horoscope etc. Though such type of pieces are countable but it is said that this idea was inherited from the Roman examples of continuous narration.[65]

In the Gandhāran reliefs, in between the figures of the Buddha, there are depiction of Bacchanalian and erotic scenes which usually do not go with Indian ideas. So it is said that it is a western influence

61. S. Paul, "Discovery of Gandhāra Sculptures" in Devendra Handa (ed.), *Recent Studies in Indology*, Vol. II, Delhi, 1989, p. 412.

62. *Ibid.*, 421.

63. V.A. Smith, *A History of Fine Arts in India and Ceylon*, Bombay, 1969, p. 49.

64. L. Nehru, *Origins of the Gandhāran Style*, Delhi, 1989, p. 17. fn.22.

65. *Ibid.*, p. 17.

on Gandhāran art. The Bacchanalian scenes - depicted on the stair-riser panels discovered at Jamālagarhi, for example, consist of several groups of two or three figures engaged in dancing, drinking, playing in musical instruments or in mildly erotic play.[66]

A few Gandhāran reliefs show the use of more space between figures and groups of figures which is said to have inherited from the west.[67] The representation of *Mahāparinirvāṇa* and the sixteen ascetics on the relief are similar to the dramatized representations on many Roman sarcophagus.[68]

Some Gandhāran Buddha images in round reflect the same realistic apprehension of reliefs as their western prototypes, but these are not the complete prototypes. Some images, like Hoti-Mardan Buddha image, are strikingly close to Western prototypes. The realistic treatment of facial features and drapery, the *dehanchment* in the figures stand, reflect the western influence.[69]

But, only on the basis of style and technique which were supposed to be employed for the carving of the Gandhāran sculptures, this school should not be termed as Graeco-Buddhist art. Even all the sculptures are not the Greek or Roman prototypes. The image or images which are very much near to Apollo type image are countable. It is notable here that the art forms and motifs of Gandhāra do not come to view before the Greek domination of this area became a thing of the past[70] and that the patrons of this school were principally the Central Asian Śakas and Kuṣāṇas. Even the technique borrowed from Hellenistic standards is modified by such trends as Iranian, Scythian, etc., but the themes depicted are Indian, almost exclusively Buddhist. Moreover, the technique too seems to be gradually Indianised. Thus, we may say that in the so called Graeco-Buddhist art, there is nothing directly associated with Greece. So the Gandhāra school should get the nomenclature of its own i.e. Gandhāra school

---

66. *Ibid.*, pp. 16-17.
67. *Ibid.*, pls.8,7a.
68. *Ibid.*, pls. 9,10,11.
69. *Ibid.*, p. 25.
70. R.C. Majumdar (Ed.), *The Age of Imperial Unity*, Bombay, 1953, p. 519.

of art. This is why A.H. Dani says that the Gandhāran art is the product of its soil.[71]

ORIGIN : There are different opinions regarding the origin of Gandhāra school of art. Foucher says that Gandhāra art owes its parentage to Hellenistic ideals practised in Roman empire from where it spread to Gandhāra and patronized by the Śakas and the Kuṣāṇas. Hellenistic elements cast their shadow before the Kuṣāṇa period and the Bimaran reliquary of the period of Azes I demonstrates the connection between Roman and Gandhāra.[72] Benjamin Roland attributes the parentage of Gandhāra school of art to Roman art and comparing it with the evolution of images of Roman art with Gandhāra art establishes his contention.[73] John Marshall gives the credit of the origin of Gandhāra art to philhellinic art of Parthians in the first century B.C.[74] Wheeler says that Roman influence was the corner stone of Gandhāra art which made its way to the Kuṣāṇa empire.[75] Kākāsu Okakura opined that Gandhāra art shows more Chinese influence than Roman. According to him Gandhāra art is influenced by the art of Hān period.[76] On the basis of French Archaeological expedition, Dobbins says that "recent discoveries in Afghanistan and Russian Central Asia have supplied the Hellenistic art of Gandhāra with a parentage."[77] The Russian excavation at Transoxiana postulates the existence of a school of art but its relationship with Gandhāra art has not been fully established.[78]

Let us examine the above mentioned opinions. First the sculptors who carved the images of the Buddha in the Gandhāran

---

71.   A.H. Dani, quoted in F. Sehrai, *The Buddha Story in Peshawar Museum*, Peshawar, 1978, p.7.

72.   Foucher, quoted in B.B. Shrivastava, *Op. Cit.*, p.337.

73.   Benjamin Rowland,*The Art and Architecture of India-Buddhist, Hindu and Jain*. London, 1953, pp. 75-100.

74.   J. Marshall, *Op. Cit.*, p.17.

75.   Wheeler, quoted in Stavisky, "Central Asia in Kuṣāṇa period" in D.P. Chattopaddyaya (Ed.), *Kuṣāṇa Studies in USSR, Indian Studies - Past and present*, Calcutta, 1970,p.50.

76.   Kākāsu Okakura, quoted in B.B. Shrivastava, *Op. Cit.*, p.337.

77.   K.W. Dobbins, *Op. Cit.*, p.55.

78.   *Ibid.*, pp. 55-56.

region should not be considered as foreigners because they were perfectly Indian. It is well known that through the centuries a considerable number of people belonging to different races i.e. the Persian, Greek, Roman, Śakas, and the Kuṣāṇas were settled down in the North-western region of India. Subsequently they intermingled with local Indian people and they came in contact with the Indian ethos. Gradually they became Indianised and were absorbed within the indigenous cultural mainstream obviously with certain modification as a result of their differing cultural pattern. But in general, they came under the influence of Buddhism.[79] The Gandhāra school of art, devoted primarily to the service of Buddhism, represents really a stage and process of this Indianization and should be viewed in that light instead of emphasising the extraneous factor.[80] Interestingly some scholars think that Greek or Roman sculptors were imported to create Gandhāra art. But A.H. Dani challenges this view and says that the classical taste was already present in the Gandhāran region since a long time. So it was not necessary to import artists from outside. If the Kuṣāṇas needed the help of the artists, they could summon the sculptors from Mathura which was not only nearer than Greece or Rome but also it was under their stronghold and a famous centre of art.[81] He is of the opinion that the Gandhāra art is the product of the indigenous sculptors.[82] It was Foucher who "imbued with sympathetic understanding for all that is best in Greek classical tradition",[83] put stress on the Greek origin of all the best sculptures. But now this theory does not hold ground.

DATING AND SPAN OF THE SCHOOL: The problem of dating and span of the Gandhāran school of art has not been unanimously solved. Cunningham says that the reign of Kaniṣka and his successors was the golden period of the Gandhāran art.[84] Generally scholars are

---

79. S.K. Saraswati, *A Survey of Indian Sculpture*, Calcutta, 1957, p.71.
80. V.K. Thakur, "A study of the contribution of Buddhism to the Gandhara School of Art", (*Proceedings of the 6th International Buddhist Conference*, Bodhagaya, Vol. VI, 1980), p. 155.
81. F. Sehrai, *Op. Cit.*, p.7
82. *Ibid*.
83. D.L. Snellgrove (Ed.),*The Image of the Buddha*, Delhi, 1978,p. 71.
84. A. Cunningham, *ASR*, Vol. III, p.39.

in agreement with Cunningham. Fergusson says that the time span of Gandhāra school stretches from the first to fifth century A.D. and its golden period was around 400 A.D.[85] Smith[86] puts the origin of this school in the Ist century A.D. and 600 A.D. being its exteme date. According to him the culmination of the Gandhāra school of art was from 50 to 150 or 200 A.D. He also considers that the reign of Kaniṣka to be its best period and the works of good quality to have been produced in the first three centuries of the Christian era.[87] Grundwedel is of opinion that second half of the 4th century A.D. was the golden period of this school of art, but Vogel says that Gandhāra school of art developed and declined in 1st century B.C. - A.D.[88] Benjamin Rowland says that the Gandhāra school of art came into existence in the last lap of 1st century A.D. and persisted till 4th century A.D.[89] J. Marshall opined that the Gandhāran art originated during the Śaka period, reached in maturity during the Kuṣāṇa period, temporarily declined in the first half of the 3rd century A.D. and flourished between the later part of the 4th century A.D. and towards the end of the 5th century A.D.[90] This phase of Gandhāran art extended over a much wider area than earlier.[91] Stylistically, Harald Ingholt has tried to establish the chronology of the Gandhāra school of art[92] but that has not been found to be correct. However, the stylistic chronology made by Ingholt are as follows:[93]

Phase one : The sculptures of this phase (144 to 240 A.D.) show Greek and Iranian influence.

Phase two : (240 to 300 A.D.) Increase in Iranian influence with Greek influence.

Phase three : (300 to 400 A.D.) Influence of Mathura school of art.

---

85. J. Fergusson, *A History of Indian and Eastern Architecture*, Delhi, 1967. pp. 181-182.
86. V.A. Smith, *Op. Cit.*, pp. 50-51.
87. *Ibid*.
88. Quoted in B.B. Shrivastava, *Op. Cit.*, p.344.
89. Benjamin Rowland, *Op. Cit.*, pp. 75- 100.
90. J. Marshall, *Op. Cit.*, pp.17,67,112,110.
91. *Ibid.*, pp. 110-112.
92. Harald Ingholt, quoted in B.B. Shrivastava. *Op. Cit.* p.344
93. *Ibid*.

Phase four : (400 to 460 A.D.) Dominence of Sassanian influence.

On the basis of stylistic index based on stratified evidence, Dobbins has structured the chronology of Gandhāra art which he has fixed by the dated sculptures,[94] though, the chronology fixed by Dobbins is almost similar to those of Marshall.

Bimaran reliquary : This relic casket was discovered at Bimaran, Afghanistan by Mason in the year 1834-37.[95] The casket is decorated with a series of arched niches with standing figures of the Buddha flanked by Indra and Brahmā. Here the Buddha is shown in standing posture with his right hand raised in *Abhaya mudrā*. The facial features and the garment is in Gandhāran style.[96] There is also a fourth image depicted on the casket in the worshipping gesture. The image is wearing the *dhoti*, as normally worn by the Bodhisattvas. It is also adorned with neclace, armlets, bracelets and ear-rings etc. The hair is dressed in a top knot. The earlobes are long. Joe Cribb [97] conjectured this image to be a Maitreya Bodhisattva, because the image is ornamented which is usually considered as Bodhisattva Maitreya. Here one thing is notable that the image does not hold the *Amrit-Kalash-Kamaṇḍala* in the left hand which we usually find with the Maitreya Bodhisattva images. Above on the triangular space between the two arches, two eagles hovering in the air, is shown. The pillars of arches bear an oblong cut. A few coins of king Azes II were found with the casket and on this basis some scholars have dated it in the period about 50 B.C. If so, this has to be considered as the earliest known representation of the Buddha in anthropomorphic form.

But the dating of the reliquary has been challenged by various scholars on various grounds. Coomaraswamy opined that "the methods of excavation nearly ninety years ago were not by any means as critical as they are now, coins in any case merely provide a *terminus*

---

94.   K.W. Dobbins, *Op. Cit.*, p.57.

95.   A.K. Coomaraswamy, *The Origin of the Buddha Image*, Delhi, 1972, p.33.

96.   K.W. Dobbins, *Op. Cit.*, fig.22.

97.   Joe Cribb,"Re-examination of the Buddha images on the Coins of King Kaniṣka...", in A.K. Narain (Ed.), *Studies in Buddhist Art of South Asia*, Delhi, 1985, p. 82.

post quem[98]..." Lohuizen-de-Leeuw [99] is of opinion that the little oblong cut on the pilasters is a late motif because the oldest Gandhāran specimens do not display this shape. The design of row of ogees (S-shaped curve) resembles the late architecture of about 3rd century A.D. as seen on the *Stūpa* of Shevaki at Kabul. Apart from these the shape of the pilasters also suggest a later date as in the earlier phase we have the round corinthian pilasters.[100] So, all these features do not support to place the Bimaran reliquary in an early period. The numismatic evidence tells us that the mound of Sirkap near Taxila was in habitation before Kaniṣka only. The coins have been found from this site is upto the period of Vima Kadaphises.[101] So, Van Lohuizen aptly says that if the Bimaran reliquary is to be dated from about 50 B.C. then why do we not get any Buddha figure till Kujula Kadaphises i.e. till the middle of the 1st century A.D.[102]

The other points which favour the early dating of this reliquary are the discovery of four copper coins of Azes II with it, its fine workmanship and inscription engraved on the Steatite vase which contained this casket. As regards the coins of Azes is concerned, it may be noted here that the date of the reliquary may not neccessarily be the same as that of the coins. It is well known that the deposite of coins with the casket was considered to be an auspious act. The coins of Azes might have been current at the time of the deposit because sometimes the old coins remains current for centuries, as at Mathura itself as R.C.Sharma says that the Kuṣāṇa coins were in circulation till the beginning of 20th century.[103] So, on the basis of coins deposit it is very difficult to date any object. Joe Cribb considered these four 'bronze' coins posthumous imitations of the coins of Azes II, because the coins bear the name of Azes, but have inscriptions, types and central marks etc. which we do not find on the regular issues of either Azes I or Azes II.[104] According to Cribb, coins of this

---

98. A.K. Commaraswamy, *Op. Cit.*, p.33.

99. J.E, Lohuizen-de-Leeuw, *The Scythian Period*, Leiden 1949, pp. 84-85.

100. R.C. Sharma, *Buddhist Art of Mathura*, Delhi, 1984, p.153.

101. *Ibid.*

102. J.E. Van Lohuizen, *Op. Cit..* p.87.

103. R.C.Sharma,*Op.Cit.*, p. 153.

104. Joe Cribb, *Op. Cit.*, p. 82.

type could have been issued from the end of the reign of Azes II upto the beginning of Vima Kadaphises.[105] Apart from these, Cribb further says that none of the coins were found within the reliquary or the pot containing it so it can not be claimed that there is any chronological relationship between them.[106] It is only possible to say that the relic chamber of the *Stūpa* was closed later than the reign of Azes II.

As regards the fine workmanship is concerned, obviously it is a debatable matter, however, Joe Cribb says that there is a close relationship between the images of the Buddha on the casket and those which appeared on the coins of Kaniṣka. Cribb also links the lotus design on the base of the casket with those of Kaniṣka's reliquary.[107]

As mentioed earlier, there is a Kharoṣṭhi inscription engraved on the Steatite Vase of the casket. But it also does not help in dating it earlier. Sten konow puts[108] it to the later part of the 1st century B.C. He says that "from the point of view of palaeography this does not seem to be any objection to a dating of the Bimaran Vase as about contemporaneous with the Mathura lion Capital."[109] Joe Cribb says that the epigraphy is typical of the Kuṣāṇa period and can not be more precisely dated.[110] Thomas on the basis of palaeography[111] favoured a date between 50 and 78 A.D. The difference of about a century indicates that this is not at all a forceful plea in favour of early dating of the reliquary. Thus on the basis of above discussion we may say that Bimaran reliquary must belong to the later half of the 1st century A.D.[112]

A Kharoṣṭhi Inscription of Senavarma, King of Oḍi:

This is a recently discovered inscription, on a painted figure of the Buddha in the Gandhāran region of the time of Kujula

105. *Ibid.*, p. 82.

106. *Ibid.*

107. *Ibid.*

108. Sten konow, *Kharoṣṭhi Inscriptions*, Vol. II, part I, Varanasi, 1969, p. 51.

109. *Ibid.*

110. Joe Cribb, *Op. Cit.*, p. 82.

111. Thomas, quoted in R.C.Sharma, *Op.Cit.*, p. 154.

112. J.E.Van Lohuizen, *Op. Cit.*, p.94;R.C.Sharma, *Op.Cit.*, p. 154.

Kadaphises. The inscription is written in North-western Prākrit and it is engraved on a gold plate. It is dated in the regnal year 14 of Senavarma who was a king of Oḍi (in the Swāt Valley), whose name is not very much known, perhaps he was a vassal of the Kuṣāṇa king Kujula Kadaphises.[113] The inscription refers to the deposition of the body-relics(*Śarira*) of the Bhagavant, (i.e. the Buddha) in the Eka-Kuṭa *Stūpa* by the king Senavarma. The relevant lines of the inscription which reads as, *"Ji'ase likhita ya sarira pra'ithavani's Sanghamitrena 'Ali'a-putrena 'anaka' ena karavita ya Sadi'ena Sacakaputrena meri akhena ukeḍa ya......"*(and the body- relics(*Śarira*) to be deposited were painted to the life by Sanghamitra son of Ali'a the *anankaios*(royal Kinsmen), and were ordered by Sadi'a son of Saccaka the meridarkhes(distt. officer...). Regarding the implication of the inscription B.N.Mukherjee says and add that " these were engraved (Ukeḍa) can only mean that the casket destined to contain the supposed body-relics of the Lord Buddha was painted with one (or more than one) life-like(and not symbolic) representation of the Master, and that it was engraved (with figure and designs)."[114] In support of his argument he mentions ,[115] " the Vases bearing inscriptions, referring to the deposition of such relics, have sometimes been found to contain pearls, beads, and other items of stone, smaller caskets, coins etc. These objects cannot be expected to have been painted. But their main container might have been, if necessary, decorated with colour. So the *Śarira*, "painted to life", as mentioned in the epigraph in question, refers to the body-relics and its or their, painted container. If the *Śarira* or rather the container part of the *Śarira* was "painted to life", the implication should be that the container was painted with one (or more than one) life-like representation of the Master(i.e. of the Master's body or *Śarira*). The other supported body-relics, such as those mentioned above, could not conceivably have been "painted

113. H.W.Bailey,"A Kharoṣṭri Inscription of Senavarma, king of Oḍi," *JRASGBI*, 1980, mp I, pp. 1ff; B.N.Mukherjee, "New light on the Kuṣāna period," *Monthly Bulletin, Asiatic Society*, Jan. 1981, pp. 11f; K.K.Dasgupta, "Origin of the Buddha Image, in *Studies in Ancient Indian History*, Delhi, 1988, p. 152.

114. B.N.Mukherjee,"Earliest Datable Iconic Representation of the Buddha," *Journal of the Varendra Research Museum*, Vol. 6, Rajshahi, pp. 11-14.

115. B.N.Mukherjee, 'Forword' in R.C.Sharma, *Op. Cit.*, pp. viii-ix.

to life", It appears that the receptacle containing the relics was deposited in the Eka-kuṭa *Stūpa* and the act was perpetuated by the record incised on the gold-plate in question." If Mukherjee's opinion is right then we may connect this one to the Bimaran reliquary where the figures of the Buddha is shown on the side.

Any way, whether Mukherjee is right or not in his argument it can not be said perfectly, but it is clear that as the inscription suggests, the Budddha figures were shaped in the Gandhāran region even before the Kaniṣka period. In this connection we may recall the tradition of the first representation of the Buddha in the pictorial medium.[116] That the images of the Buddha were made before the Kaniṣka period is further evidenced by the images of the Buddha (in relief) discovered at Butkara I, Pakistan. Tha images have been placed between the period circa first century B.C.- A.D.[117]

Medium of Art: The material which were used in the Gandhāran school of art were metal, terracotta, stucco and stone.[118] Metal was usually used for religious Utensils such as reliquary etc. and terrracotta for the manufacture of popular images. Stucco was more sophisticated and permanent material than terracotta as an art medium and seems to have been popular among the monastic communities.[119] Stucco was also easily available and economical.[120] Stone was also abundantly used for images, steles and reliefs and it seems to be popular among the rich patrons of the religious faith. The different types of stones were used such as phyllite, pale grey micaceous schist, chloritized micaceous schist, hornblende schist, quartz schist, talcose chloritic schist and steatite etc.[121] Phyllite stone, although poor in quality than other stones, was more popular mainly ·due to its easy availability.[122] The locality also influenced the selection of material, such as black phyllite was available in the Peshawar Valley.

---

116. K.K.Dasgupta, *Op. Cit.*, p. 152.
117. D. Faccena,'Excavations of the Italian Archaeological Mission(ISMEO) in Pakistan: Some Problems of Gandhāra Art and Architecture, "*Central Asia in the Kushana period*, Vol. I, Moscow, 1975, pp. 150 f.
118. K.W.Dobbins, *Op. Cit.*, p. 57.
119. *Ibid.*
120. *Ibid.*
121. J.Marshall,*Op.Cit.*, p. 65.
122. *Ibid.*, pp. 65-66.

3.4.   Iconography of the Buddha images:

The chronological study of the images of the Buddha from the Gandhāran school of art is very difficult due to the lack of datable images etc. Though, there are a number of inscribed images but they are countable and even those which bear inscriptions, mention certain specific era but whose period has not been unanimously fixed up by scholars in terms of Christian era.[123] Marshall says that "Gandhāra school comprised of various groups of atelier, located in different parts of the country, each with its own traditions and its own individual style.[124] According to Dobbins[125] the study of the Gandhāran sculptures can be based on the stratified evidence only. The earliest known phase of Gandhāran art has been traced in the Bhir Mound and Sirkap sites at Taxila which are divided into three strata. The earliest one is Bhir Mound III. The second perod is traceable at Bhir Mound II and Sirkap VII. The third perod is revealed in materials from strata VI and V. The art finds firstly show the local and central Indian and then Iranian, Hellenistic and Roman features.

The general features of the Gandhāran Buddha images manifest that the seated or standing image is conceived as a short, rather stocky, and the position of the body is invariably frontal, wherever he appears independently. The hair is arranged in waves gathered together on the top of his head. The eyes are open and there is a little circle (*Ūrṇā*) between the eyebrows. He usually wears a moustache. The ear-lobes are distended. The body is covered in a heavy monastic cloak that hangs in deep folds and covers the standing figures to just above the feet, while the seated figures may be covered completely. The hands are usually held in a gesture of *Abhaya*, *Dharmacakra-pravartana* etc. The Buddha is usually seated on lotus throne.[126],

---

123. K.W. Dobbins, *Op. Cit.*, p. 58; N.G. Majumdar, *A Guide to the sculptures in Indian Museum*, part II, ASI, Delhi, 1937, p.18; V.S. Agrawala, *Indian Art*, Varanasi, 1965, pp. 271f.

124. J. Marshall, *Op. Cit.*, p.17.

125. K.W.Dobbins, *Op. Cit.*

126. D.L.Snellgrove (Ed) *Op.Cit,67*; K.K. Dasgupta, *Op. Cit.*, *pp.149-150*

The images of the Buddha can be divided broadly into three sections:A,B and C.

SECTION A: In this section those images of the Buddha will be discussed which are approximately assignable to the period of first to 4th century A.D.

The images of the Buddha discovered at Butkara I (Pakistan) belonging to the period first century B.C.-A.D. are now considered as the earliest representations of the Buddha from the Gandhāran side. These sculptures are depicted in a group of panels, dislplaying, *inter alia*, the figures of the Buddha.[127] Stylistically these figures are slightly different from those of the other general Gandhāran sculptures. It seems that these images indicate the local influence on it. Here the head of the images of the Buddha are round. The *Uṣṇiṣa* is very much prominent. The facial expression of the images are usually charming. In the standing images, the drapery is similar to other Gandhāran style, however, there are some variations in the folds of the garment. The other variations can be noticed in the hair style.

The images of the Buddha depicted on the Bimaran reliquary, once which were considered as the earliest representation of the Master in the Gandhāra region, have been already discussed in detail.

The next stage of development is traced in the group of reliefs from Sikri.[128] Here we notice certain stylistic change in the images of the Buddha, e.g. changes can be noticed in the 'boxed end' pleats, which become less conspicuous and the hem of the seated images of the Buddha from Shaikhan Dheri II became almost straight and which only slightly shows the 'box'.

During the first century A.D. again some changes in the style of the images of the Buddha are traceable. Now the sculptures are marked by sturdy and natural and increased focus on the central figure surrounded by lesser figures. After 1st century A.D., particularly in the 2nd century A.D., we notice certain changes in the style. In the seated images the Buddha's right shoulder is bare and in the standing images, the right elbow is held away from the body. After

---

127. D. Faccena, *Op. Cit.*, pp. 150f. fig. 8,44,51,52.
128. J. Marshall, *Op. Cit.*, pls. 49-53.

the Kuṣāṇa period the images of the Buddha reflect the Sassanian influence. One of the features of this influence is 'wind-blown' treatment of drapery.[129]

Some of the specimens of the images of the Buddha which bear above-mentioned features are as follows:

(1)   A standing image of the Buddha housed in the National Museum of Pakistan at Karachi is an important specimen of this section.[130] Here the Buddha is shown standing with an ascetic. The forearms of both the hands are broken, however, the gesture of the remaining portion of the right hand shows that previously it might have been turned in *Abhaya mudrā*. The heavy folded drapery, which covers both the shoulders, is similar to the typical known Gandhāra style. The hair is wavy with prominent *Uṣnīṣa* and the rounded halo is plain but slightly damaged from the side. There is *Ūrṇā* between the eyebrows. The Buddha also wears the moustache which is clearly visible.

(2)   A seated image of the Buddha, discovered at Jamalgarhi and housed in the Indian Museum Calcutta (No.334) is another representative of this section.[131] Here the Buddha is shown seated on throne. Four eglantine flowers are depicted on the front face of the throne. The forearms of both the hands are missing. The halo is decorated at the edge by a zigzag line and is slightly inclined to front. The hair is treated in wavy curls and above the forehead, runs in parallel horizontal lines. The folded drapery covers both the shoulders of the figure. The minute face and the slim body of the image is noteworthy. The image is assigned to the period circa 3rd century A.D.

(3)   The next representative of this group is a seated image of the Buddha housed in the Indian Museum Calcutta (No.336). [132] Here the Buddha is shown seated on a box-shaped throne. The forearm of the right hand is broken and the gesture of the

129. K.W. Dobbins, *Op. Cit.*, pp. 72-76.
130. D.L. Snellgrove,*Op. Cit.*, pl.41.
131. N.G. Majumdar, *Op. Cit..*, pl.II,b.
132. *Ibid.*,pl.IIa,p.62.

remaining portion indicates that it might have been raised in *Abhaya mudrā*. The folded drapery covers the left shoulder only and the right shoulder is bare. The treatment of hair has to be marked, specially the protuberant portion of the knot which is made highly ornamental. The rounded halo is plain. The face of the image is small and slightly turned down towards the front side. The front of the pedestal bears two Bodhisattva figures in meditation, of whom one is holding a flask who would be identified as Maitreya. Between the figures there is a depiction of a miniature scene of the Indraśāla cave of Magadha. Inside the cave the Buddha is seated in meditative posture. To the Buddha's left, God Śakra is standing with folded hands and to His right, the heavenly musician Panchasikha is playing on the lyre. The feature of the image is not very much impressive.

(4)   The next representative of this section is a preaching image of the Buddha discovered at Sahr-i-Bahlol, housed in the Archaeological Museum, Peshawar.[133] Here the Buddha is shown seated on a throne and hands are turned in preaching posture. The garment which covers the left shoulder only hangs in the heavy Gandhāran style. The rounded halo is plain. The hair is wavy with prominent *Uṣṇiṣa* and there is *Ūrṇā* or circular mark between the eyebrows. The image having strong shoulders which perhaps indicate the strength. The next notable feature of this image is the lion motif on the throne which we usually do not find with the Gandhāran Buddha image. There is depiction of a small seated figure on the throne in meditative posture with halo which is identified as Bodhisattva. The Bodhisattva is flanked by two standing figures with folded hands who may be identified as the worshippers.

(5)   A standing image of the Buddha, housed in the National Museum, New Delhi(Pl.6), is another representative of this section.[134] This image of the Buddha is a fine example of the typical Gandhāran style. Here the Buddha is shown standing

---

133. D.L. Snellgrove, *Op. Cit.*, pl.42.
134. *Ibid.*, pl.43.

on lotus throne on which two small figures are depicted with folded hands. The forearm of the right hand is broken, however, the remaining portion indicates that it was raised in *Abhaya mudrā*. The hair is arranged in waves gathered together on the top of his head. The eyes are open and there is a circular mark or *Ūrṇā* between the eyebrows. The rounded halo is plain. The earlobes are distended. The body is covered with a heavy monastic cloak that hangs in deep folds and covers the body upto just above the feet.

(6)   A seated image of the Buddha housed in the National Museum, New Delhi(Acc. No. 68.187) is the next representative of this section. Here the Buddha is shown seated on a throne with his right hand in *Abhaya mudrā* (Pl.4). The fingers setting of the right hand looks like as webbed fingers. The folded and heavy drapery covers both the shoulders. The earlobes are long but slightly damaged at the lower portion.   The eyes are open and the circular mark or *Ūrṇā* between the eyebrows that we usually find in the Gandhāran Buddha image, here it is conspicuously absent. The hair is arranged in waves with promenent *Uṣniṣa*. Here the circular halo is absent. The face of the image is rather round in comparison to other Buddha images belonging to this section.

(7)   The next representative of this section is a standing image of the Buddha housed in the National Museum, New Delhi (Acc. No. 49.21/1,Pl.9). Here the Buddha is shown standing but his throne is absent. The folded drapery covers both the shoulders; here the folds of the drapery is not as deep as we have noticed in other images of this section. The forearm of the right hand is broken, however, the gesture of the remaining portion indicates that it was turned in *Abhaya mudrā*. The hair is arranged in the wavy manner with prominent *Uṣniṣa*. The eyes are open and there is circular mark between the eyebrows. The earlobes are distended. Here the nose is sharp. There are round marks on the neck. The circular halo is decorated at the edge.

(8)   A standing image of the Buddha housed in the National Museum, New Delhi (No.48.3/43,Pl.7) is another representative of this section. Here the Budda is shown standing on a lotus throne. He wears the heavy and folded drapery which covers both the shoulders. Here again the forearm of the right hand is broken, however, it seems that it was turned in *Abhaya mudrā*. The circular halo is plain. The hair is arranged in wavy manner with prominent *Uṣniṣa*. Here the sharp nose is slightly damaged on the tip. The eyes are open and looks slightly upward. The *Ūrṇā* between the eyebrows is either absent or damaged as the spot shows.

(9)   A pilaster housed in the National Museum, New Delhi(Acc. No. 48.3/32,Pl.8) shows the standing image of the Buddha on each side. Here the image is in relief which shows that the Buddha is standing on a throne which is partially damaged. The right hand of the Buddha is turned in *Abhaya mudrā*. The remaining features are similar to other Gandhāran Buddha image as mentioned above.

(10)   A seated image of the Buddha in relief housed in the National Museum, New Delhi(Acc. No. 483/38,Pl.10) is another representative of this section. Here the Buddha is shown seated on a box like throne and surrounded by worshippers. The images are slightly damaged. The right hand of the Buddha is turned in *Abhaya mudrā*. Here the change can be noticed that in this image as the *Uṣniṣa* has been made in top-knot manner as we find in some of the images from Mathura. Here the halo is absent, remaining features of this image are similar to other Gandhāran sculptures as mentioned above, though, the front portion of the face is damaged.

(11)   A relief panel depicting the Buddha in a goat-cart, housed in the Victoria and Albert Museum, London, is another representative of this section.[135] Here the Buddha is shown seated

---

135. L. Ashton (Ed.), *The Art of India and Pakistan*, London, 1947-8, p.36,pl.19.

in a goat-cart drawn by two rams, the reins being held by a male figure in the centre. Behind the Buddha there is a monk who is followed by a second male attendant, above them there is a tree. To the left of the tree there is a male figure carrying a box-like object in the left hand. To the left again there are three male figures, two of them are carrying small boxes in the left hand and boards with handles possibly for writing, in the right hand.

(12)   The next representative of this section is also a relief panel, housed in the central Museum, Lahore.[136] Here the panel depicts three episodes related to the life of the Buddha. On the right, Sumāgadhā, the daughter of Anāthapiṇḍaka, is seen stripped to the waist and chastising the naked ascetics at the house of her prospective father-in-law. The house is represented by a pillared platform, and on the top are seated her fiance's parents. The father is bearded and looks angry, his right arm being extended as though intervening in the girl's struggle with the ascetics. Immediately above the figure of Sumāgadhā in the first scene, she is again represented, this time turning her back on the house and perhaps appealing to the Buddha for his intervention. In the centre, in a third scene, the Buddha is seen decending from air. On the Buddha's right, Sumāgadhā is seen kneeling in adoration and behind her there is a group of six male and female figures paying homage to the Buddha. Actually this story is depicted in the *Divyāvadāna*[137] which will be further discussed in the proper place.

(13)   A standing Buddha image in bronze, housed in the Victoria and Albert Museum, London, is the next representative of this section.[138] Here the Buddha is shown standing with his right hand in *Abhaya mudrā*. The left hand is holding the hem of the garment. The hair is wavy with prominent *Uṣṇiṣa*. The drapery hangs in loose folds, covering both the shoulders. The pedestal

---

136. *Ibid.*, p.37, pl.20.
137. B.N. Chaudhury, *Op. Cit.*, p. 195.
138. L. Ashton, *Op. Cit.*, p.39, pl.20.

is missing, but the image retains its *Prabhā torana* which has an ornamental fringe.

(14)   A seated image of the Buddha in terracotta, housed in the Provincial Museum, Lucknow, is the next representative of this section.[139] Here the Buddha is shown seated in *Padmāsana*. The forearms of both the hands are missing. The folded drapery covers both the shoulders. The hair is arranged as usual in wavy manner with prominent *Uṣṇiṣa*. The pedestal depicts the rosetts, deer and elephants.

SECTION - B : As discussed earlier, only a few Gandhāran sculptures are dated. So, it is very difficult to place them in proper chronological sequence. However, it is believed that the changes in the application of the image are the same in Gandhāra as in Mathura school of art. Actually after the fall of the Kuṣāna empire, the Sassanids made their influence over the North-western region of India[140] which affected the Gandhāran art too. The Hindukush became a kind of cultural watereshed. On the eastern side (Nothern Pakistan) and Kabul Valley, the influence of Indian style began to predominate. On the Western side, in Fonduskistan etc. the Persian influence appears especially in the dress and decorative ornaments. But, even so, Buddhist art remains basically of Indian inspiration.[141]

A large number of images, particularly the heads of the Buddha have been discovered belonging to the later Gandhāran art which reflect a different conception of the images of the Buddha from the earlier ones. Though the features remain the same, such as the hair, arranged in schematic waves of separate locks, the same type of *Uṣṇiṣa* and *Ūrṇā*, the sharp chiselling of the eyes, lips and nose etc. But the difference is in general treatment like, the lines are softer and the chiselling half-closed eyes shows a mellow and spiritual disposition. "It is possible that these later Gandhāran images fall short of the mature Indian ascetic vision simply because the whole style was limited from the start by Graeco-Roman human proportions and standards of beauty. The abstraction of an intellectually idealized vision, which

---

139. *Ibid.*, p. 39.

140. R.C. Majumdar (Ed.), *Op. Cit.*, p. 132.

141. D.L. Snellgrove (Ed.), *Op. Cit.*, pp.102-3, pl.66.

transcends altogether the perception of the physical form, was never a fundamental part of their tradition. Even though they were clearly aware of the ideal of supreme Enlightenment as expressed in the Indian classical traditions of the Gupta period, the Gandhāran craftsmen were prevented by the tradition already received from elsewhere from rendering the idea perfectly in concrete plastic form. Their Buddha faces doubtless suggest compassion and sublime knowledge, as well as the serene inwardness of yogic experience"....[142]

Some of the specimens of the images of the Buddha which bear above mentioned features are as follows:

(1)   A seated image of the Buddha housed in the Kabul Museum is a representative of this section.[143] Here the Buddha, depicted in relief, is shown seated on a throne but in a typical manner. He is surrounded by devotees in elegant Sassanian-style costumes. There are two large male figures which stand in graceful attendance on either side. Here it is notable that the Buddha on which He is shown seated is the Indian-style arch. So, we may say that this representation is a short of synthesis of both Sassanian and Indian styles.

(2)   The next representative of this section is a seated image of the Buddha discovered at Loriyan Tangai and housed in the Indian Museum, Calcutta (No. 274).[144] Here the Buddha is seated on a lotus throne. His dreamy, half-closed eyes are specially to be noted. The circular halo is plain. The wavy hair is tied into a knot on the top of the head. The flowing locks of hair separating from the middle of the forehead. There is a circular mark or *Ūrṇā* between the eyebrows. The hands of the Buddha are turned in preaching pose. The drapery, which covers the left shoulder only, shows a definite attempt to make it close-fitting to the body. The image is finely modelled, having a distinct softness which reminds us the ideals of the Gupta sculptures.

---

142. *Ibid.*, p. 107.
143. *Ibid.*,pl.66.
144. N.G. Majumdar, *Op. Cit.*, pl. frontispiece.

(3)   The next representative of this section is also a seated image (Pl.5). Here the Buddha is shown in the preaching pose and seated on the pedestal of a lotus seat flanked by two Bodhisattvas and two tiny human figures who may be identified as the devotees. The folded. garment covers the left shoulder only. The hair is wavy with prominent *Uṣniṣa* and there is *Ūrṇā* in between the eyebrows. The remaining features are similar as in the case of general Gandhāran images.

SECTION - C : The iconography of Gandhāran art in Afghanistan belonging to the later period differs in certain points, though not in general.

At Ai Khānum, the city in northern Afghanistan on the bank of river Oxus, a centre of Greek artistic culture of Bactria has been traced out. Some of the works found there beloning to the 3rd century B.C. are supposed to have been produced in Greece itself, but there are also some pieces like heads both in stucco and in unbaked clay, which indicates a stylistic evolution towards distinctive new forms, although they are still definitely 'classical'.[145] The implantation of Greek motifs in the distant regions of Bactria and the territory beyond the Oxus is also traced by the unbaked clay sculptures of Khalchayan which belongs to the period circa first century B.C. and Soviet archaeologist G.A. Pugachenkova considered it one of the first manifestations commissioned by the Kuṣāṇas.[146]

There is a great difference in style between the clay sculptures discovered in Ṭāpā Sardār near Ghazni and the stucco sculptures from Haḍḍā. Though this phenomenon was not only confined to Haḍḍā only, the stucco sculptures from Haḍḍā are wholly Gandhāran whereas the clay one from Ṭāpā Sardār and from Haḍḍā itself are classical Hellenistic elements.[147]

---

145. D.L. Snellgrove, *Op. Cit.*, p. 178.
146. G.A.Pugachenkova, quoted in Ibid., 178.
147. D.L. Snellgrove, *Op. Cit.*, pl. 132.

Some of the specimens of the images of the Buddha which bear above mentioned features are as follows:

(1)   A seated image of the Buddha from Ṭāpā Shotor (Haḍḍā) is the first representative of this section.[148] In the niche of a temple courtyard at Ṭāpā Shotor (Haḍḍā), the Buddha is shown seated with Vajrapāṇi who is shown seated with his thunder bolt (*Vajra*). Here we can see what a faithful transposition of a classical prototype of the figure of Vajrapāṇi in the form of *Heracles*.

(2)   The Kabul Museum housed a beautiful Buddha head, discovered at Ṭāpā Kālān, Haḍḍā.[149] The head is the finest example produced by the late Gandhāran school. Here we notice the synthesis of both Hellenistic and Indian ideals as reflected in the figure e.g., the form of the eyes, clearly marked planes of the eyebrows, and the shape of the lips. Other specially iconographic elements, such as the *Ūrṇā*, the *Uṣniṣa* and the elongated earlobes are, of course, Indian.

---

148. *Ibid.*
149. *Ibid.*, pl. 133.

CHAPTER - 4
ICONOGRAPHY OF THE BUDDHA IMAGES IN SĀRNĀTH SCHOOL OF ART

4.1 : *SĀRNĀTH : Historical perspective of the place.*

4.2 : *Buddhism in Sārnāth.*

4.3 : *Sārnāth School of Buddhist Art.*

4.4 : *Iconography of the Buddha images.*

4.1 : SĀRNĀTH : Historical perspective of the place:

Sārnāth, which is situated near modern Benaras in Uttar Pradesh, occupies a very prominent place among the Buddhist centres. This is the place where the Buddha had delivered his first Sermon, which is known as *Dhammacakkapavattana*, to the *Panchvaggiyas* (viz, Aññāta-koṇḍañña, Bhaddiya, Vappa, Mahānāma and Assaji). This is also the place where the Buddhist Saṅgha was formed for the first time.[1]

In Buddhist literature Sārnāth is known as Ṛsipatna and Mṛgadāva or Mṛigadāya. The word Ṛsipatana means the abodes of Ṛsis. In the *Mahāvastu*,[2] the place was called Ṛsipatana, as it was here the five hundred *Pratyeka Buddhas* or *Ṛsis* had attained *Nirvāṇa*. According to other Buddhist source, Isipatana (Pāli) was so named because Sages on their way through the air, got down here and started from here on their aerial flight (*isayo ettha nipatanti uppatanti cati- Isipatanam*).[3] In the *Divyāvadāna* the place is called *Rshivadana*[4]

---

1. N.Dutta, and K.D.Bajpai, *Development of Buddhism in Uttar Pradesh*, Lucknow, 1956, p. 33.

2. E.Senart, *Le Mahāvastu*, I, Paris, 1882, pp. 357ff.

3. G.P.Malalasekera, *DPP*, Vol. I, p. 324.

4. N. Dutta & K.D.Bajpai, *Op. Cit.*, p. 332.

which reading is also found in the Chinese works. The other name of the place is Migadāya or Migadāva (Skt. Mṛgadāya or Mṛgadāva), because it was a roaming ground of deers.[5] The name 'Sārnāth' seems not to be very old. According to General Cunningham it was formerly the name of a local Śiva temple. He drives this name from Sāraṅganāth (lord of deer) a name of Lord Śiva.[6] Still there is a temple nearby in which the image of Mahādeva is enshrined.[7]

So far as the early history of Sārnāth is concerned, during the age of *Mahājanapadas* it was the part of Kāshi *Mahājanapadas*. So, the anecdote of Sārnāth was related to Kāshi. Later on, Sārnāth along with Kāshi was included into the Mauryan empire. Aśoka got erected here a big *Stūpa*, remnants of which are still present, known as *Dharmarājikā Stūpa*. A monolithic Aśokan pillar adorned with four lion figures and surmounted by *Dharmacakra* was also erected on the spot where the Buddha had delivered his first Sermon.[8]

After the fall of the Mauryan empire, Sārnāth was the part of the Śuṅgan (2nd-1st cent. B.C.), the Kuṣāṇas(1- 180 A.D.), and the Gupta-Harṣa(320-650 A.D.) empires respectively. After Harṣa, Sārnāth came under the Gurjara-Pratihāra rulers of Kanauj and after them, Sārnāth came under the sway of the Pāla rulers of Bengal.[9]

4.2　Buddhism in Sārnāth

Sārnāth is considered as the birth place of Buddhism. As stated above this is the place where Lord Buddha had delivered his first Sermon. The *Mahāparinibbāna Suttanta of Dīgha Nikāya* mentioned it as one of the four places of pilgrimage for the devotees of the Buddha.

After the demise of the Buddha and before the rise of Aśoka, the history of Buddhism in Sārnāth is not very much clear. During the Aśokan period, Sārnāth became a famous centre of Buddhism as is evidenced by the Buddhist remains found there.

5.　B.N.Choudhury, *Buddhist Centres in Ancient India*, Calcutta, 1982, p.66.

6.　N.Dutta & K.D. Bajpai, *Op.Cit.* p. 332.

7.　V.S.Agrawala, *Sarnath*, ASI, New Delhi, 1980, p.2.

8.　N.Dutta & K. D. Bajpai, *Op. Cit.*, pp.333-334.

9.　*Ibid.*, pp. 334-335.

The *Mahāvaṁsa* mentions that there was a large community of Buddhist monks at Sārnāth (Isipatana) in the second century B.C.[10] It further mentions that twelve thousand monks from Sārnāth under the leadership of Dhammasena attended the foundation ceremony of the *Mahāthūpa* in Anurādhapur.[11]

The prevalence of Buddhism in the second century B.C. is further corroborated by a dozen railing pillars belonging to this period which have been discovered near the *Dharmarājikā Stūpa.*[12]

Buddhism was prevalent during the Kuṣāna period as evidenced by the discovery of several Buddhist images belonging to this period and place.[13]

During the Gupta period, Sārnāth became a great centre of Buddhism and art. The Chinese pilgrim Fa-Hien who visited the place in early fifth century A.D. mentions that he had seen many *Stūpas* and two monasteries at Sārnāth.[14]

Sārnāth also received the patronage of king Harṣavardhana of Kanauj. Another Chinese traveller Huien- Tsang, who visited the place during the Harṣa period, has left a graphic account about the prevalence of Buddhism at Sārnāth. He mentions that he had seen at Sārnāth fifteen hundred monks belonging to the Sammitiya school of Buddhism.[15]

Buddhism further continued to flourish at Sārnāth during the reign of the Pāla kings and lingered upto the 13th century A.D.

4.3.   Sārnāth school of Buddhist art

Sārnāth school of Buddhist art has prominent place in the history of Indian art; it has its own identity. The history of this school of art can be traced back from the days of Aśoka. Aśoka, who is well known as a great patron of Buddhism, got erected several

---

10. *Mahāvaṁsa* XXIX, 31.

11. *Ibid.*

12. V.S. Agrawala, *Op. Cit.,* p. 4.

13. N.Dutta & K.D.Bajpai,*Op.Cit,* p.334.

14. James Legge, *A Record of Fa-Hien's travels...,* Oxford, 1886, p. 94.

15. Thomas Watters, *On Yuan Chwang's Travels in India,* Vol. II, London,1904-05, pp. 48-54.

monuments at Sārnāth. The *Dharmarājikā stūpa*, the fragments of which have been discovered, is believed to be erected by Aśoka.[16] Another important monument of Sārnāth is the monolithic pillar which is inscribed with Aśoka's schism edict that threatened those monks and nuns whoever will bring schism in the Saṅgha will be expelled from the congregation (Brotherhood).[17'] The pillar has on its surface magnificent polish which was once surmounted by the magnificent lion- capital and now this is adopted by the government of India as her state emblem. Now the detached capital is preserved in the local Museum. The capital was crowned by a large *Dharmacakra*, several fragments of which have been recovered. [18] The portion of the capital consists four components,(i) a lotus at the base, (ii) a circular abacus carved with an elephant, bull, horse and lion separated from one another by a wheel each, (iii) four lions set back to back over the abacus, and (iv) a surmounting wheel of which fragments, have been discovered,[19] as stated earlier. In later period two more records were inscribed on this pillar, one of the Kuṣāṇa period which referred to the reign of Aśvaghoṣa, a rular of Kausambi, and the other one of the early Gupta period which mentions the teachers of the Sammitiya sect.[20]

During the Śuṅgan period (2nd-1st century B.C.) a railing was constructed most probably round the Aśokan *Stūpa*, the fragments of which have been discovered and now is preserved in the local Museum.[21] On these railings we notice the worship of the Buddhist symbols viz, *Stūpa*, *Dharmacakra*, *Triratna* etc.[22] Apart from this, a Yakṣa image and a female image showing in grief, have also been discovered at Sārnāth, which is noticeable, belonging to the Śuṅgan period. The sculptures of Sārnāth of the Śuṅgan period show the same flatness and round form of the face as in the art of Sānchi,

---

16. D.Mitra, *Buddhist Monuments*, Calcutta, 1971, p. 67.

17. D.C.Sircar, *Select Inscriptions*, Vol. I, Calcutta, 1942, p. 75.

18. D.Mitra, *Op. Cit.*, p. 66.

19. *Ibid*.

20. *Ibid.*, p. 67.

21. N.Dutta and K.D.Bajpai, *Op. Cit.*,p. 334.

22. *Ibid.*, p.392.

Bharhut and Mathura etc.[23] Some broken sculptures bearing the shining polish have also been discovered at Sārnāth.[24] Among the other monuments of the Sungan period is a crescent-shaped temple of which only the foundation is now extant.[25]

During the Kuṣāṇa period, Sārnāth became a full- fledged school of art. Several images of the Buddha and Bodhisattva have been discovered at Sārnāth belonging to this period. The early images of Sārnāth are mainly based on the ideals of Mathura school of art. One of the finest examples of this period is a colossal image of Bodhisattva in red sandstone donated by friar Bala.[26] The image is inscribed in Brāhmi script and its language is mixed Sanskrit and Prākrit. The inscription records that this image was set up by monk Bala in the 3rd regnal year of king Kaniṣka.

The inscription further mentions about the contribution of the Buddhist monks to the faith. The friar Bala is also credited for the donation of similar type of images at Mathura and Śrāvasti. In the 3rd and 4th century A.D. there seems to be a decline in the output of the Sārnāth school and the images of the Post-Kuṣāna period are very few. But in the 5th century A.D., Sārnāth school suddenly sprung into action and showed a tremendous increase in plastic actively. It is notable here that the Sārnāth school of art is independent from the Gandhāran influence; it derives its technique from the Mathura school and further improves the Mathura traits and gave it the highest expression possible. Commenting on Sārnāth sculptures, Stella Kramrisch says that "the Śārnāth version of the Mathura prototype is subtler than the original".[27] The decline of Mathura school placed the Sārnāth school of art at the top among artistic centres. It influenced the style of later Mathura sculptures and other provincial art centres. But the influence of Sārnāth school was more widely felt towards the east than to the west.[28]

---

23. *Ibid.*, p. 393.

24. *Ibid.* p. 334.

25. *Ibid.*

26. *Ibid.*, pp. 393-394.

27. Stella Kramrisch, *Indian Sculpture*, New Delhi, 1981, p. 63.

28. R.D.Banerjee, *The Age of the Imperial Guptas*, Delhi, 1981, p. 168.

Apart from the images in round, we also find several slabs from Sārnāth which depicts the events of life of the Buddha and the *Jātaka* stories etc.[29]

In the post Gupta and the Pāla period onwards we find the rise of Vajrayāna tradition of Buddhism which gave a new dimension to Buddhist art. Several Buddhist deities came into existence and their images were carved out. However, the images of the Buddha had still maintained their significance, but their number had been reduced.

The images of the Sārnāth school have been praised by all and compared to be the best Buddha images which are far outclassed whereas the highest point reached by the Mathura school was completely mature and manifests expression of the consciousness, but in Sārnāth it was the expression of bliss.[30] S.K.Saraswati aptly says ," In the majority of creations, Sārnāth maintains a hightend intellectual and aesthetic consciousness of more than ordinary interest. The softened plasticity and the sublime spiritual grace constitute the striking features of the Sārnāth ideom and in Gupta classicism this idiom plays the most conspicuous part. A gradual emphasis on slender form and refined contours ultimately leads to an almost weightless physiognomy that seems to have soared above all earthly moorings."[31]

4.4   Iconography of the Buddha Images:

The anthropomorphic images of the Buddha came into existence at Sārnāth school of art in the Kaniṣka period, but those images were supposed to be imported from Mathura school.[32] We have hardly any evidence that the images of the Buddha and Bodhisattva were being made in Sārnāth in the post-Maurya and pre-Gupta period epoch.[33] However, the Sārnāth Museum possesses some Buddhist images of the Kuṣāṇa-Mathura style made of Chunar sandstone which lend support to the possibility of earlier sculptural activities of Sārnāth.[34]

---

29.   N.Duttta and K.D.Bajpai, *Op. Cit.*, p. 396.

30.   S.K.Saraswati, *A Survey of Indian Sculpture*, Calcutta, 1957, pp. 134-136.

31.   *Ibid.*, p. 136.

32.   R.D.Banerjee, *Op. Cit.*, p. 166.

33.   R.C.Sharma, *Buddhist Art of Mathura*, Delhi, 1984, p. 240.

34.   R.D.Banerjee, *Op. Cit.*, p.166.

During the Gupta period we find a break-through in the evolution of Sārnāth school of art which emerged as a blooming centre of plastic art and produced some of the most magnificent and graceful images of the Buddha. However, the Sārnāth school follows the graceful technique which may be considered as a further improvement of the Mathura trait.[35]

The characteristic features of the images of the Buddha of Sārnāth school of art.[36]

(1)   The face of the images are oval and the heads are covered with short curly hairs with *Uṣṇiṣa*.

(2)   The eyebrows are long and arched and the eyes are down cast and more than half closed, covered with long upper eyelid. The nose is sharp. The lips are full and the lower lip is slightly drooping. The chin is strong. The neck has double fold imparting a feeling of inner strength. On the whole, the facial expression conveys calmness, serenity and deep spiritual expression.

(3)   The body and limbs of the images are slender and supple. The standing images are relaxed with slightly bent knees imparting a certain feeling of litheness and movement.

(4)   The drapery is foldless and reduced in volume which closely follow the body contour like a transparent sheet. The indication of drapery only survives in the thin lines on the body suggesting the frills and edges of the garment. The sides falling apart are given as filmsy muslin like texture. The drapery covers both the shoulders making a semicircle round the neck and the garment end is held in the left hand. The drapery spreads like a fan near the feet of the seated images. Though the foldless wet drapery is usually regarded as a distinguishing feature of the Sārnāth style as mentioned above, but recently a few images of the Buddha discovered from the site of Govindnagar,

35. R.C.Sharma,*Op. Cit.*, p. 241.
36. N.Dutta and K.D.Bajpai, *Op. Cit.*, pp.394-395; S.K.Saraswati, *Op. Cit.*, pp. 134-136.

Mathura, proves that this idiom was also invented by the Mathura school.[37]

(5)   The centre of the halo is plain and surrounded by concentric bands of beads and vegetal motifs. It is simpler than the halo of the Mathura school. In a few cases the halo covered the whole body and is almond shaped or a simpler plain slab without halo is also seen in the images of late 5th century A.D. The seated images are shown in *Vajraparyaṅka* on cushion.

(6)   The wheel surrounded by a pair of deer and devotees is depicted on the pedestal of the seated images which signify the first preaching of the Buddha at Sārnāth.

(7)   In Sārnāth school various *mudrās* or hand poses are shown viz., *Abhaya*, *Varada*, and *Dharmacakra* etc. They also show sometimes webbed fingers (*Jālāṅguli*).

(8)   Here the images other than the Buddha are tiny or absent or given place on the pedestal.

(9)   The overall impression of the images conveys peace, serenity, calmness and deep spiritualism. The softened plasticity and the sublime spiritual grace constitute the striking features of the Sārnāth idiom. The simplicity, less decorative motifs, gliding contour line and felicity of expression make the images of Sārnāth school graceful and spiritual. Stylistically the images of the Buddha of Sārnāth school can be divided into three sections:

SECTION-A : In this section those images of the Buddha will be discusssed which are assignable to the period of 5th century A.D.or before.

The images of the Buddha belonging to this period mostly inherits the Mathura traits and some of the images are supposed to be a Mathura production and set up at Sārnāth. The poses of the images are stiff and straight with broad shoulders and prominent chest. But the images which belong to the period about 4th century A.D. manifest heavyness and clumsiness etc. Usually the right hand of the images of the Buddha is turned in *Abhaya mudrā* and the left hand held akimbo resting on the waist. But in some images we find

---

37.   R.C.Sharma, *Op. Cit.*, pp. 218,224,241.

that the left hand is resting slightly below the waist. The drapery is usually transparent and covering the left shoulder only and it is slightly folded. The lower garment is fastened by a waist band which terminates into a double knot and hangs down on the right thigh and the garment further reaches below the knee and its hem rests on the left hand.

Some of the specimens of the images of the Buddha which bear above mentioned features are as follows:

1. A standing image of the Buddha discovered at Sārnāth and assigned to the period 2nd-3rd century A.D. is the first representative of the section. The image has been carved according to the Kuṣāṇo-Mathura idiom. The facial expression and other features are mainly same as those discussed above.[38]

2. A headless standing image of the Buddha discovered at Sārnāth is another representative of this section.[39] The image is made of chunar sandstone. The right hand of the image is turned in *Abhaya mudrā*. Here notable thing is that the palm of the right hand is slightly turned towards the right side from the body. The fingers are made in webbed manner but not in manner of Maṅkuār Buddha image. The left hand with clenched fist supporting the heavy and conventional folds of the outer robe which follows the idiom of Mathura. But the physiognomy of this image had shed much of the heavy and clumsy weight of its Mathura lineage and has grown taller and slender in limbs. The drapery which covers the left shoulder only has been treated integrally with the plastic treatment of the body surface except at the folded ends. The head and the halo of the image are missing. This image is said to have marked an intermediate stage before the 5th century transition.

SECTION .- B : The images of the Buddha from the Sārnāth school belonging to the period 5th-6th century, manifest certain slylistic changes. Now we find that the belt is worn around the waist of the

---

38. J.G. Williams, *The Art of Gupta India - Empire and Provinces*, New Delhi, 1983, p.6.

39. D.L. Snellgrove, *The Image of the Buddha*, Delhi 1978, pl.61, p.101.

image.[40] The eyebrows defined by a raised ridge, seem to be the early example of this group.[41] Now the halo of the images do not follow a consistent pattern like Mathura and seems a matter of individual variety.[42] Some images show only a mark around the waist and is more slender than the belted one.[43]

Some of the specimens of the images of the Buddha which bear above mentioned features are as follows:

1.   A standing image of the Buddha discovered at Sārnāth, housed in the National Museum, New Delhi (Acc. No.59.527/2,Pl.11) and belonging to the period 5th century A.D. is a representative of this section. The forearm of the right hand and the wrist of left hand are broken, however, the remaining portion of the right hand shows that it was turned in *Abhaya mudrā*. The head is oval-shaped and the hair is arranged in curly manner with prominent *Uṣṇiṣa*. The eyes are half closed and the earlobes are elongated. The foldless drapery covers both the shoulders making a semicircle round the neck. The end of the garment seems to be held by the broken hand. Around the waist there is a belt. The lower portion of the legs are broken. On the whole, the facial features of the image convey calmness, serenity etc.

2.   Another representative of this sectinon is a standing image of the Buddha, housed in the National Museum, New Delhi (Acc.No.59.527/3,pl.12). The image is well preserved in comparison to the previous one. Here the Budha is shown standing on a simple plain pedestal. The body of this Buddha is slender in comparison to the earlier one. The head is also oval-shaped but slightly changed from the previous one. The hair is arranged in curly manner but in a vertical order whereas the previous one was horizontal. The *Uṣṇiṣa* is as usual prominent. The eyes are half closed and the earlobes are elongated. Here

---

40.   J.G. Wiliams, *Op. Cit.*, p. 76.

41.   *Ibid.*. pls.85,86,87.

42.   *Ibid.*, p.17.

43.   *Ibid.*, pl.89.

also there are lines marked round the neck. The foldless drapery covers both the shoulders and closely follow the body. The end of the gartment is held by the left hand. The right hand is turned in *Abhaya mudrā*. Here the halo covers the whole body and is almond-shaped. The edge of the halo is decorated whereas the remaining portion is plain. There is a mark round the waist.

3.   Another representative of this section is a standing image of the Buddha which is housed in the National Museum, New Delhi (Acc.No.49.116,Pl.13). The features of this image are more or less similar to the previous one. Here the Buddha is flanked by two tiny figures. The figure depicted on the right side is holding something like *Vajra* on his right hand so it may be Vajrapāṇi and the figure depicted on the left side looks like as if it is adorned with ornaments so it may be a Bodhisattva. Here it is notable that both the tiny figures are very much rubbed so nothing can be said with certainty, even the Buddha figure is also rubbed. The foldless drapery covers both the shoulders. The right hand is turned in *Abhaya mudrā* and the left hand is holding the end of the garment. Here the halo is a simple piece. The notable thing is that here the umbrella has been depicted above the head and which is attached to the plain halo. Here also there is a mark round the waist.

4.   Next representative of this section is a seated image of the Buddha, housed in the National Museum, New Delhi (Acc.No.47.21,Pl.14). Here the Buddha is shown seated on a pedestal and his hands are turned in *Dharmacakrapravartana mudrā*. Here again the Buddha is flanked by two male figures, those standing on the right side holding, most probably a *Vajra* on his right hand while the gesture of the image depicted on the left side is not clear. On the pedestal, six human images are depicted. Since the images are very much rubbed, their facial features can not be stated. In the centre a wheel is depicted. Here also the halo is replaced by a plain back piece.

5.   The next representative of this section is a seated image of the Buddha, housed in the Sārnāth Museum. This image is in the

pose of depicting the first Sermon of the Buddha. Here also the Buddha is shown seated on a pedestal. The oval head of the Buddha is covered with small curly hairs which is within the halo. Here the eyebrows meet with nose to form three angles of 120 degrees.[44] The well decorated halo is carved with a pair of celestial figures and conventionalized floral scroll work. The eyes are half closed. There is compassion and sublimity all round. On the pedestal, figures of the five monks are depicted whom the Buddha had given his first Sermon. There is also a figure of a woman with a child at the left corner who may be a donor. In the centre, the wheel is depicted which is being worshipped by the monks. This is considered as one of the best works[45] of art of not only the Sārnāth school but the whole of India also.

6.   A standing image of the Buddha belonging to the period 6th century A.D., housed in the Sārnāth Museum is the next representative of this section.[46] Here the Buddha is shown in *Varada mudrā* and his drapery suggests to be a later production. The halo is also here replaced by a large almond-shaped back piece which is larger than the image and is plain except its border which has a beaded and scalloped ring all round it.

SECTION-C: At Sārnāth, we find that specific events from the life of the Buddha were depicted independently or collectively on a number of bas-reliefs and steles. There are a number of steles which depict four or eight main events of the life of the Buddha on a single stele in different combination.[47] Later on it was adopted by the Eastern school of art in post-Gupta and Pāla period on a large scale.[48]

---

44. J.C. Williams,*Op.Cit.*, p.78, pl.94.

45. R.C. Majumdar (Ed.), *The Classical Age*, Bombay, 1954, pp.519-520; D.L. Snellgrove, *Op.Cit.*, pl.63; N. Dutta and K.D. Bajpai, *Op.Cit.*, p.395;V.S. Agrawala, *Op. Cit.*, p.27, pl.VIII.

46. D.L. Snellgrove, *Op.Cit.*, p. 100, pl.64.

47. Ratan Parimoo,*Life of Buddha in Indian Sculptures*, New Delhi, 1982, figs.89-95.

48. *Ibid.*, pp. 58-72.

Some of the specimens of the images of the Buddha which bear above mentioned features:

1.   The first representative of this section is a Stele housed in the Sārnāth Museum(No.C(a)1) which shows the scenes from the Budddha's life in four panels viz., (i) Birth, (ii) Enlightenment, (iii) first Sermon, and (iv) Decease.[49]

2.   The next representative of this section is also a stele housed in the Sārnāth Museum (No.C(9)3). The stele depicts the eight events of the life of the Buddha. The four principal ones are the same as mentioned above, these are carved on the four corners of the slab. The remaining scenes illustrate the presentation of honey by a monkey to the Buddha, the Buddha taming the mad elephant Nalagiri, the Buddha's decent from the heaven of the Thirty three gods and the great miracle of Śrāvasti when the Buddha multiplied his form in thousand fold.[50]

---

49. *Ibid*
50.   V.S.Agrawala, *Op. Cit.*, pl.IXA.

CHAPTER - 5

ICONOGRAPHY OF THE BUDDHA IMAGES IN KASHMIR SCHOOL OF ART

5.1 : *KASHMIR : Historical perspective of the place*

5.2 : *Buddhism in Kashmir*

5.3 : *Kashmir school of Buddhist art*

5.4 : *Iconography of the Buddha images*

5.1 : KASHMIR : Historical perspective of the place.

The ancient Kingdom of Kashmir seems to have been a part of the present state of Jammu and Kashmir. It was geographically adjacent to Gandhāra and for sometimes Kashmir was the constituent part of Gandhāra. In the Buddhist texts both the countries are mentioned together.[1] But the Chinese traveller Huien-Tsang distinguished Kashmir from Gandhāra and dealt with the two countries separately.[2]

The political history of Kashmir before the rise of the Mauryan King Aśoka is obscure. Though there are textual references to certain political activities but most of them are not corroborated with historical facts. However, from the Aśokan period onwards the political history of Kashmir is more or less clear. Kalhaṇ,[3] the twelfth century chronicler of Kashmir, says in his work *Rājataraṅgini* that Aśoka, in course of his religious activities, who ruled over the place, founded the city of Srinagar and built a large number of *Stūpas* at Sushkaletra and

---

1. *Milindapañha*(Ed.), V.Trenckner, p. 331; *Mahāvaṁsa*, XII,13-25; H.C. Raychaudhury, *Political History of Ancient India*, Calcutta, 1950, p. 308; B.N.Chaudhury, *Buddhist centres in Ancient India*, Calcutta, 1982, p. 156.

2. T.Watters, *On Yuan Chwang*, Vol.I, London, 1904-05, p. 26.

3. *Rājataraṅgini*, Vol.I., Eng.tr.M.A.Stein, Delhi 1961, Vol.I, Verse 104.

Vitāstātra(Vithavutur). But Kalhan did not elaborate about the political activities of Aśoka in Kashmir. He mentions that Aśoka had left behind his son Jalauka to rule over Kashmir.[4] It is said that Jalauka was not a follower of Buddhism, rather to a certain extent anti-Buddhist and he persuaded a number of Brāhmanas of Kanauj to settle in Kashmir. He was a powerful ruler. He had succeeded in subduing the Greeks who had pushed their way earlier in Kashmir during the Aśokan period. After the fall of the Mauryan empire the Indo-Greeks, the Śakas, the Pārthians and the Kusānas extended their sway over most of the parts of the North-western regions of India respectively, so Kashmir naturally came under their powerful leadership.[5] There are evidences regarding the activities of the Kusāna kings in Kashmir, viz., Huska, Juska and Kaniska who are said to have laid the foundation of three towns as, Ushkur, Zukur and Kaneshpur.[6] They also built many *Caityas*.[7] One of the greatest religious events in the Kusāna period was the convention of the fourth Buddhist council under the patronage of Kaniska in which Buddhism was broadly divided into two sections: Hīnayāna and Mahāyāna i.e. conservative and liberal sects of the Buddhism, respectively.

After the fall of the Kusāna empire, Kashmir came in the hands of a king known as Abhimanyu. He was succeeded by one Gonanda who is said have laid the foundation of the Gonanda dynasty. It is said that he was a great patron of the Nāga- cult in Kashmir.[8]

The establishment of the Huna rule in Kashmir in the last quarter of the fifth century A.D. proved to be another land mark in the political history of Kashmir. Toramāna, the Huna chief had ruled over Kashmir in the last quarter of the fifth century A.D. Toramāna was succeeded by his son Mihirkula in about 515 A.D.[9] Mihirkula is known as an anti-Buddhist. He extended his royal patronage to the

---

4.   Ibid., Verses-115-200; B. Malla, *Sculptures of Kashmir*, Delhi, 1990, p.6.

5.   S.Chattopadhaya, *Early History of North India*,Calcutta, 1958, pp. 60-61.

6.   D.Mitra, *Buddhist Monuments*, Calcutta, 1971, p. 107;
     R.K.Parmu, *A History of Muslim Rule in Kashmir*, Delhi, 1969, p. 54.

7.   *Ibid*.

8.   R.K.Parmu, *Op. Cit.*, p. 54; B. Malla, *Op.Cit.*, p.6.

9.   P.N.K.Bamazai, *A History of Kashmir*, Delhi, 1962, pp. 74-77; R.K.Parmu, *Op. Cit.*, pp. 55-56.

Brāhmaṇas in Kashmir and helped them to revive Brāhmaṇical customs and traditions.

The history of Kashmir, after the death of Mihirakula and before the rise of the Kārkoṭa dynasty, witnessed the rise and fall of at least twenty-five kings. Most of them were insignificant and only three out of the twenty-five kings received certain amount of political prominence. Gopāditya is to be mentioned as first who was responsible for the foundation of Gupkar area of Srinagar.[10] The second prominent ruler was Mātrigupta, who is said to have acquired the throne of Kashmir with the help of the king Vikramāditya of Ujjain.[11] Pravarasena was the third prominent ruler who also laid out the city of Srinagar.[12] Balāditya was the last ruler of the Gonanda dynasty.

After the decline of the Gonanda dynasty in Kashmir, Kārkoṭās came into power. The first ruler of the Kārkoṭa dynasty was, Durlabhavardhana who had married the daughter of the last Gonanda king, Balāditya, who had no son.[13] During his reign Huien Tsang visited Kashmir and he found a well deposed and hospitable king in Durlabhavardhana. Huien-Tsang mentions that Takṣaśilā, Siṁhapur, Urasā, Pan-nu-tso and Rajpura states were under Kashmir.[14] Thus Durlabhavardhana ruled not only over Kashmir proper, but over a part of the Western and North-western Panjab as well. He gave patronage to Buddhism. However nothing of historical importance is known of Durlabhavardhana. The chronology of the Kārkoṭa dynasty as mentioned by Kalhaṇa has been a subject of much controversy. He mentions the name of kings and assigned them to different dates following the Lauka era but he did not mention clearly about the dates of the accession of a number of kings of Kārkoṭa dynasty. He however, mentioned the period of a number of kings. Durlabhavardhana was succeeded by his son Durlabhaka who ruled for fifty

---

10.   B.Malla, *Op. Cit.*, p.7.

11.   *Rājatarangini*, Verse-242 and introduction, p. 83.

12.   *Ibid.* iii, 339-49.

13.   R.C.Majumdar (Ed.), *The Classical Age*, Bombay, 1954, p. 132.

14.   S.Beal, *The life of Huien-Tsang*, New Delhi, 1973, p. 68, fn.192; R.C.Majumdar, *Op. Cit.*

years.[15] Durlabhaka was succeeded by his eldest son Chandrāpiḍa who ruled for a short period of eight years and eight months.

Chandrāpiḍa had to face an Arab invasion during his reign which was led by Muhammad Bin Qasim and the situation turned so precarious that in 713 A.D., the king sent an envoy to the Chinese emperor asking for help against the Arabs. But he did not get any help from China. However, he was able to defend his kingdom against Arab aggression and drove them beyond his frontiers.[16] Chandrāpiḍa was renowned for his piety and justice. He was succeeded by his brother Tārāpiḍa, an oppressive rulers who became the enemy of humanity. His inglorious rule of fours years was full of cruel and bloody deeds. He was followed by his younger brother Lalitāditya Muktāpiḍa, the celeberated king of Kashmir who is known to have ruled over Kashmir between 724-761 A.D. He was eager for conquests and passed his life mainly on expeditions.[17] He is considered as a great conqueror like Alexander the great. He had defeated Yasovarman of Kanauj. By that victory, Lalitaditya acquired the right of suzerainty over the larger area. According to kalhaṇa after defeating Yaśovarman, Lalitāditya proceeded to the eastern ocean and reached Kaliṅga.[18] Kalhaṇa says that his conquests included Karnataka, Konkan, Dvārakā and Avanti. He is known to have fought with the king of Gauḍa, but it is also said that Gauḍa king probably had acknowledged his suzerainty without a fight, for that he sent elephants to join the army of Lalitāditya.[19] His Himālayan invasion included the subjugation of Kamboja, Tukhāras, Daras, Bhauttas and Mummuni respectively. He had also conquered Jalandhara and Lohara. Lalitāditya is said to have carried his army to Uttarakurus and its adjoining areas.[20] Lalitāditya

---

15.   S.C.Ray, *Early History and Culture of Kashmir*, Calcutta, 1957, pp. 46-47; R.C.Majumdar (Ed.)., *Op.Cit.*, p. 132.

16.   A.K.Majumdar, *Concise History of Ancient India*, Vol. I, New Delhi, 1977, p. 301; R.C.Majumdar (Ed.)., *Op. Cit.*, p. 132.

17.   J.Narain and R.Tiwary, "Lalitāditya Muktāpiḍa, The first Antiquarian King of India", in K.K.Dasgupta & others (Ed.), *Studies in Ancient Indian History*, Delhi, 1988, p. 194.

18.   R.C.Majumdar, *Op. Cit.*, p. 134.

19.   *Ibid.*

20.   G.L.Kaul, *Kashmir Through the Ages*, Srinagar, 1963, p. 29.

proceeded in his conquering expedition as far as Bengal in the east through Banda, Vāranasi, and Monghyr which is recorded.[21] His coins have been discovered at Banda, Rājghāt and Monghyr.[22] Kalhaṇa also mentions that after conquering Kanauj, Bengal and Kalinga, Lalitāditya reached the banks of Kaveri and conquered the Deccan or South India.[23] But R.C.Majumdar opined that it is difficult to believe without corroborative evidence that Lalitāditya had conquered South India.[24] From Deccan Lalitāditya marched towards Dvārakā region.[25] According to A.K. Majumdar, Lalitāditya could not annex Gauḍa to his kingdom.[26] It is said that he had also conquered the Sahis.[27] Lalitāditya had sent a diplomatic mission in 733 A.D. to the Chinese emperor in order to induce him against the Tibetans.[28] It is further said that the mission was received with honour by the emperor who recognised the King of Kashmir as his royal ally, but no military assistance was sent from China, nevertheless, Lalitāditya succeeded on his own in defeating not only the Tibetans but also the mountain tribes on the north and North-Western frontier of his kingdom, such as the Dards, Kambojas and Turks.

During the reign of Lalitāditya Kashmir flourised to a great extent. The ruins of the Mārtanda temple still form the most striking remains which have survived the ancient architecture of Kashmir.

Lalitāditya was followed by his two sons Kuvalyapiḍa and Bajrāditya for a short period. Again two insignificant kings namely, Prithvipiḍa and Sangrāmapiḍa ruled for a brief period and then Lalitāditya's grandson Jayapiḍa (son of Bajrāditya) became the king. Like Laltāditya he had also undertaken a campaign in the plains of

------

21. K.S.Saxena, "Kuṣāna Rule in Kashmir," *Bulletin of Museum and Archaeology* in U.P., No.3, 1969, p. 38; R.C.Majumdar, *Op. Cit.*, p. 134.

22. B.Malla, *Op. Cit.*, p. 8.

23. *Rājatarangini*, Vol.I, iv, 131.

24. R.C.Majumdar, *Op.Cit.*, p. 134.

25. K.S.Saxena, "The Deccan Compaign of Lalitāditya of Kashmir", *Bhārtiya Vidyā*, Vol.38, 1978, p.67.

26. A.K.Majumdar, *Op. Cit.*, p. 302; R.C.Kak, *Handbook of the Archaeological and Numismatic Sections of the S.P.S.Museum*, Srinagar, Calcutta, 1923, p.4.

27. K.S.Saxena,"Shahi- Kashmir Relations, *Bhārtiya Vidyā*, Vol. XXVII, 1967, p. 8.

28. R.K.Parmu, *Op.Cit*, p. 57; R.C.Majumdar, *Op. Cit.*, p. 133.

India and he is known to have annexed Gauḍa after defeating the king of Kanauj. But there is no adquate evidence regarding his campaign.[29] However, Jayapiḍa is said to have undertaken his campaign upto Prayāga[30] (Modern Allahabad) and from Prayāga he took an adventurous journey to Bengal in disguise and on his way, he married the daughter of a Prince Jayanta. Jayāpiḍa and Jayanta then conqured certain neighbouring areas.[31] Jayāpiḍa is credited to have defeated Bhimsena of Eastern India and and Aramudi of Nepal. It is also said that while returning from Bengal, Jayāpiḍa defeated Vajrāyudha, the king of Kanauj[32] and when he reached his kingdom after about three years of absence his throne was captured by his brother-in-law, Jajja. Jayāpiḍa had to wage war with him to get back his throne.[33] It is said that his court was embellished with scholars and poets like Kṣira, Bhaṭṭa, Dāmodaragupta etc.

Jayāpiḍa was succeeded by his son Lalitāpiḍa who is said to have misused the royal treasury for his pleasures. After him a few weak successors ascended the throne of Kashmir who could not be able to control the administration and the entire political situation was grave and disturbed. The distant territories which were captured by Lalitāditya muktāpiḍa declared themselves free from the yoke of Kashmir empire. Any how, the Kārkoṭā dynasty continued to rule over Kashmir till about the middle of the ninth century A.D.[34]

5.2 : Buddhism in Kashmir

It is very difficult to say the exact date of the introduction of Buddhism in Kashmir. However, most of the scholars held the view that Buddhism first came to Kashmir at the time of the Mauryan king Aśoka. But there is also a view that Buddhism was already prevalent in Kashmir valley before the time of Aśoka.[35] In his time also, Kashmir was known for its catholicity and toleration. It is said that some

---

29. S.C. Ray, *Op. Cit.*, pp. 44-46.
30. P.N.K. Bamzai, *Op. Cit.*, p.119.
31. *Ibid.*
32. V.A. Smith, *Early History of India,*, Oxford, 1924,p.307.
33. P.N.K. Bamzai, *Op. Cit.*, p. 119; S.C. Ray, Op. Cit., p.42.
34. R.C. Majumdar, *Op. Cit.*, p.136.
35. J.N. Ganhar, "Buddhism in Kashmir," *The Mahabodhi*, Vol.67, 1959, p.148.

Buddhist *Vihāras* were erected during the reign of a native king Surendra who lived sometime before Aśoka.[36] Kalhaṇa has clearly mentioned the erection of two *Vihāras* by Surendra.[37] One of these was at Saurasa corresponding to the present day village of Sowur on Anchar lake to the north of Srinagar. The other was at Sauraka, near the country of Dards, thus, it indicates the spread of the faith beyond the Zojila. About Surendra it is said that he was a staunch follower of Buddhism.[38]

Buddhism in Kashmir and elsewhere in India received a great impetus during the reign of Aśoka.

The *Mahāvaṁsa* mentions that after the completion of the third Buddhist council which was held at Pāṭaliputra, thirteen Buddhist missionaries were sent to different parts of the country and abroad to spread the Dhamma; among them one missionary was sent to the Kashmir-Gandhāra region under the leadership of Majjhantika.[39] Huien-Tsang and the Tibetan work *Dulvā* mention that Majjhantika was successful in converting a large number of people to Buddhism in Kashmir.[40] It is also said that Majjhantika is known to have found twelve *Vihāras* in existence on his arrival in Kashmir.[41] Kalhaṇa says that Aśoka had built a large number of *Stūpas* in Kashmir.[42] Some of the Aśokan *Stūpas* were in existence as late as the seventh century A.D. when Huien-Tsang visited the place, as mentioned by him.[43] Huien-Tsang also mentions about certain *Stūpas* in Kashmir that enshrined the relics of the Buddha.[44] The *Divyāvadāna* says that in the third Buddhist council Aśoka extended invitations to several monks of Tāmasvana in Kashmir and according to Tāranāth, Aśoka donated

---

36. J.N. Ganhar, "Buddhism in Kashmir," *The Maha Bodhi*, Vol.67, 1959, p.148.
37. *Ibid.*
38. *Ibid.*
39. *Mahāvaṁsa*, XII, 3; S. Sengupta, *Buddhism in the Classical age*, Delhi, 1985, p.18; J. Naudou, *Buddhists of Kashmir*, New Delhi, 1980, p.2; F.M. Hussain, *Buddhist Kashmir*, New Delhi, 1973, p.12.
40. T. Watters, *On Yuan Chwang*, Vol.I, London, 1904-05, p.267.
41. J.N. Ganhar, *Op. Cit.*, p.149.
42. *Rājatarangini*, Vol.I, I.101-03.
43. S. Beal, *Buddhist Records of the Western World*, Vol.I, New Delhi, 1973, pp.150-51.
44. T. Watters, *Op. Cit.*, p.261.

gifts for the maintenance of several Saṅghas in Kashmir.[45] It is said that the Indo-Greek rulers who ruled over Kashmir after the Mauryas also patronised Buddhism. They erected *Stūpas* and *Vihāras* etc. [46] The reign of Kuṣāṇa king Kaniṣka proved to be another landmark in the history Buddhism in Kashmir. The fourth Buddhist council was held at Kashmir under the patronage and presidentship of Kaniṣka and Vasumitra respectively. [47] It brought some of the most outstanding scholars and monks of the time to Kashmir which henceforth became one of the most important centres of Buddhist learning and culture.

In this council five hundred monks were called upon to write commentaries on all three *Piṭakas*. The scholars wrote commentaries on the three *Piṭakas* of 100,000 *ślokas* each. Kalhaṇa says that Nāgārjuna, the great exponent of Mahāyāna philosophy, resided for sometime in Shaḍarhaṭvana (modern Harwān).[48] According to the Chinese source large number of Buddhist scholars used to live in Kashmir during the Kuṣāṇa period viz.,, Aśvaghoṣa, Vasumitra, Vasubandhu, Dharmatrāta etc. The remnants of various *Stūpas*, *Vihāras* and number of terracotta figures of Buddha, Bodhisattva etc. discovered at Harwān and Ushkur.[49] These archaeological remains are dated mostly between the 3rd and 5th century A.D. The Buddhist scholars of Kashmir not only propagated Buddhism in the valley but did send misssionaries in the neighbouring area such as central Asia, Tibet and China.[50]

In Kashmir, Buddhism received much patronage from the Kuṣāṇas. But at the time of king Abhimanyu Buddhism did not get much support and Brāhmanism re-established their position. Again at the time of king Nara(circa 4th-5th century A.D.) whose capital was situated in the vicinity of Bijjāvihāra, Buddhism received a severe blow.[51] A large number of Buddhist edifices etc. were destroyed.

---

45. D. P. Chattopadhyaya (Ed.), *Tāranātha, History of Buddhism in India*, Simla, 1970, p.65.

46. F.M.Hussain, *Op. Cit.*, p. 13; S.C.Ray, *Op.Cit.*, p. 143.

47. According to other tradition this council was held at Jalandhara, Cf. P.V.Bapat(Ed.), *2500 Years of Buddhism,,* New Delhi, 1956,p.42.

48. *Rājatarangini*, I, 173-177

49. R.C.Kak, Ancient monuments of Kashmir, Calcutta, 1923, pp. 105-24.

50. *Ibid.*, p. 81.

51. J.N.Ganhar, *Op. Cit.* p. 150; *Rājatarangini*, I,1,199-200.

Buddhism suffered equally during the reign of the Huna king Mihirakula.[52] Mihirakula was a Śaivite and is known to have destroyed many Buddhist establishments. Both Huien-Tsang and Kalhana considered Mihirkula as anti-Buddhist and projected him as destroyer of Buddhist monastic establishments, but P.G. Pal opined that if Mihirkula was anti-Buddhist then why did he allow the Buddhist edifices of Harwān to flourish which was obviously a growing centre of Buddhist art when Mihirkula was very much in Kashmir. According to him Mihirkula was not anti-Buddhist.[53] The discovery of the Gilgit manuscripts from within a *Stūpa* testified the continued popularity of Buddhism in the post-Gupta period.[54] These manuscripts are written in character of 5th- 6th centuries A.D. and supposed to be the earliest ones so far discovered in India and were upto now unknown to have existed except in their Tibetan translations.[55] Several Buddhist sculptures which belong to 7th century A.D. are discovered from a village Pandrethān[56] (Puranadhisthāna). Apart from these, the ruins of several *Stūpas* and monasteries are also discovered from Kashmir which belong to this period. Thus, we may surmise that Mihirakula was tolerant towards Buddhism.

Buddhism again received patronage in Kashmir during the reign of Meghavāhana, successor of Mihirakula. It is said that Meghavāhana was a staunch follower of Buddhism. He had abolished the slaughter of animals and of birds and even fish. His reign is known for Buddhist foundation. The king himself built a number of monasteries and his queen Amṛtaprabhā and his other wives also patronised Buddhism and stopped animal sacrifice.[57] With the death of Meghavāhana Buddhism once again lost the royal patronage in Kashmir. But there is no record anywhere of any persecution of Buddhist monks, followers or of any discrimination towards Buddhism in the time of his successors. In fact, a number of Buddhist edifices were put up in this period. The well-known Jayendra *Vihāra* was erected at the

---

52. T.Watters, *Op. Cit.*, p. 289.
53. P.G.Pal, *Early Sculptures of Kashmir*, Leiden, 1980, pp. 63-66.
54. B.K.Deambi, Kaul, *Corpus of Sarada Inscriptions of Kashmir*, Delhi, 1982, p. 16.
55. S.Sengupta,*Op. Cit.*, p. 19.
56. S.C.Ray, *Op. Cit.*, p. 145; R.C.Kak, *Op.Cit.*, pp. 27-38.
57. *Rājatarangini*, I.iii, 7-14.

time of Pravarasena, the founder of present Srinagar who ruled in the later part of the sixth century A.D. It is said that though the Kārkoṭās of Kashmir were Hindu by faith but gave their patronage to Buddhism as well. Kalhaṇa says that Anangalekhā, the queen of Durlabhavardhana, built a Buddhist *Vihāra* which was designated as Anangabhavana *Vihāra*.[58] Kalhaṇa further mentions that Lalitāditya Muktāpiḍa(699-735) had founded the *Rājavihāra* and erected a big *Stūpa* at Parihāsapura(modern Paraspor).[59] He had also donated a huge copper image of the Buddha and patronised another *Vihāra* with a *Stūpa* at Huskapura near Baramula.[60] It is also referred that at Parihāsapura, Cankuna Tukhara, the minister of the king erected the Cankuna *Vihāra*, *Stūpa* and placed three golden images of the Buddha.[61] The same minister had also built a second *Vihāra* along with a *Stūpa* at Srinagar is recorded. It is said that in this *Vihāra* the minister installed an image of Buddha *Sugata* which was brought from Magadha on the shoulders of an elephant.[62] Kalhaṇa further says that Isanacandra (medical attendant), son-in-law of Cankuna also founded a *Vihāra*. Kalhaṇa also says that a king of Lāṭa country had erected a *Vihāra* in Kashmir which was called Kayya *Vihāra*.[63]

Jayāpiḍa Vinayāditya, another great ruler of the Kārkoṭā dynasty adorned his newly erected town Jayapura with three images of the Buddha and a *Vihāra*.[64] The Archaeological discoveries at Parihāsapura and at certain other areas testified to the flourishing condition of Buddhism during the reign of the Kārkoṭās.[65] The popularity of Buddhism reached to such an extent that the Spanda and Pratyabhijña schools of Śaivism in Kashmir to a certain extent accepted certain notions of the Buddhists.[66] However, after 8th century onwards Buddhism gradually lost its foot-holds.[67]

---

58.   *Ibid.*, I.iv.3.
59.   *Ibid.*, 194,204.
60.   *Ibid.*, 203,188.
61.   *Ibid.*, 211.
62.   *Ibid.*, 213,259-62.
63.   *Ibid*, 210,216.
64.   *ibid.*, 506-07; S.C.Ray, *Op.Cit.*, p. 146.
65.   R.C.Kak, *Op. Cit.*, p. 146.52.
66.   R.C.Majumdar(Ed.), *The Struggle for empire*, Bombay, 1979, p. 421.
67.   N.Dutta(ed.) *Gilgit Manuscript*, Vol.I, Srinagar, 1939, Introduction, p. 45.

In Kashmir Sarvāstivāda school of the Hīnayāna tradition of Buddhism flourished from the early period of its advent and its later history is depicted by literary as well as archaeological evidences.[68]

5.3 Kashmir school of Buddhist art

Kashmir school of Buddhist art produced a number of beautiful images of the Buddha, Bodhisattva etc. from early medieval period onwards. Regarding the origin or beginning of this school of art, scholars are not unamimous. The theory of A.K. Narain indicates that the history of Buddhist art in Kashmir can be traced back from the early centuries of the Christian era since the first images of the Buddha and Bodhisattva evolved in Kashmir,[69] in that particular period. It is well known that the Sarvāstivāda school which was closely allied with the Theravādins was separated as an independent school after the third Buddhist council and established themselves in Mathura, Kashmir and Gandhāran region. In course of time Kashmir became the principal seat of the Sarvāstivādins where their doctrine was taught in its purity and gradually developed into an elaborate system known as Vaibhāṣika.[70] From Kashmir the Sarvāstivāda tradition of Buddhism spread to Central Asia and adjoining areas. The Sarvāstivādins advocated the belief that every thing exists (*Sarvaṁ asti*) which is said to have become an ideological basis for the emergence of the Buddha/Bodhisattva image.

It is said that since both Aśoka and Menānder patronised the Theravāda school of Buddhism in which the Buddha was conceived as a human being, a teacher, this is why the image of the Buddha or Bodhisattva could not evolve during their period. But when the Śakas came in power who ruled over Kashmir including other adjoining areas needed to strengthen the legitimization of their rule and the Sarvāstivādins needed royal patronage for spreading their ideology of realism. Subsequently the Śakas and the Sarvāstivādins joined hands which gave birth to the first image of the Buddha/

68. S.Sengupta, *Op. Cit.*, p. 18.

69. A.K. Narain, "First images of the Buddha and Bodhisattva: Ideology and Chronology," in *Studies in Buddhist Art of South Asia*, Delhi, 1985, pp.6-7.

70. N. Dutta, *Aspects of Mahāyāna Buddhism and its relation to Hīnayāna*, London, 1930, pp.23ff.

Bodhisattva appeared on the coins of Śaka king Maues (circa 95-75 B.C.), though the seated cross-legged figure is very much rubbed and it is not clearly visible.

The Chinese pilgrim Fa-Hien[71] who visited the region, records the local tradition that the image of Maitreya was executed in wood about three hundred years after the demise of the Buddha; Huien-Tsang[72] confirms the account of Fa-Hian.

So far as the coin of Maues is concerned, nothing can be said precisely because as stated above, the coin is very much rubbed and most of the scholars do not consider it as the figure of the Buddha or Boddhisattva. The Buddhist images discovered in Kashmir mostly belong to the early medieval period. So, there is a big gap between the excavated Buddha images and the so called seated cross-legged human figure. As regards the Chinese travellers' account is concerned, though, it is obvious that if such type of image had been carved out, it can not last for a long period since the life of wood is not as long as stone. Again one question arises if the tradition of the making of the images of the Buddha or Bodhisattva existed earlier, then why did it not continue?

In Medieval period Kashmir yielded a number of Buddhist images. These images are quite significant from the iconographical point of view. Most of the images depict the Buddha in different poses. Apart from the Buddha images, the images of Bodhisattva, Śakti and the Dhyāni Buddhas etc. have also been discovered.

The features of the images show that previously it was partially influenced by the ideals of both Mathura and Gandhāra schools of art but gradually it evolved its own ideals or style.

5.4 : Iconography of the Budha images

As is mentioned earlier, the images of the Buddha or Bodhisattva were influenced by both Mathura and Gandhāra schools of art. These two traditions existed side by side in the portrayal of the garment which might cling closely to the body in the manner of the Mathura images and most of the Gupta images or they might be arranged in

---

71.   A.K. Narain, *Op. Cit.*, p. 239.
72.   T.Watters, *Op. Cit.*, p.239.

heavy fold, often covering both shoulders in the manner typical of the earlier Buddha images in Gandhāra. Similarly, the lion throne represents a Mathura tradition while the lotus throne is more likely to have been of Gandhāran provenance. Even so, combined with the lion motif on the throne there is sometimes a small *Atlas* figure, clearly of Graeco-Roman origin. Apart from this, the elephant, the horse, the peacock and the Garuḍa etc. were associated with the Buddha images. It is evident that the use of such symbols was not yet stereotyped which suggests a certain freedom of style readily associated with the disturbed social condition of Kashmir.[73]

Some of the specimens of the images of the Buddha belonging to the period upto 8th century A.D.

(1)   An image of the Buddha discovered at Pandrethan and housed in the S.P.S. Museum, Srinagar (Acc.No. 1853,pl.16) is a representative of this period. The image is in fact a mutilated bust which shows the Buddha in a standing postures.[74]Here the right hand is turned in *Abhaya mudrā* while the left hand holds the hem of the garment. The drapery covers the left shoulder only and the right shoulder is bare. The ears are elongated and the hair is arranged in curly manner with prominent *Uṣniṣa*. There are spiral lines on the neck.

(2)   A seated image of the Buddha discovered at Pandrethan, housed in the S.P.S. Museum (Acc. No.1851) is one of the best examples of the early images of the Buddha from Kashmir. The iconographic feature of this image is similar to the above mentioned image. It only displays the different hand postures. The Buddha is represented here in *Dhyāna mudrā*.[75]

(3)   A standing image of the Buddha discovered at Parihāsapura, housed in the S.P.S. Museum (Acc. No.1897) and belonging to the period of 8th century A.D. is another representative image of Kashmir belonging to the early period. Here the image bears

---

73.   D.L. Snellgrove, *The Image of the Buddha*, Delhi, 1978, pp.334- 335.

74.   R.C. Sharma, *Buddhist Art of Mathura*, Delhi, 1984, p.168; B. Malla, *Op. Cit.*, fig.6.

75.   P.G. Pal, *Early sculptures of Kashmir*, Leiden, 1986, pp. 148-149, pl.75.

certain new features. The crown and other ornaments of this Buddha-image represents him as Rājacakravarti. The image is more interesting in the sense that it bears an inscription to state that it was made in the year six of Mahārājadhirāja Parmesvara Nandi Vikramāditya. Here the Buddha appears as a king and he is adorned with Jewellery though he is shown in a monk's dress. He is shown seated on a lion throne in cross-legged posture. The right hand of the Buddha is shown blessing the king, Nandi Vikramāditya, while his left hand holds the hem of the garment. The identification of this image as the Buddha was probably done on the basis of some earlier Buddha images in the same posture found at Sārnāth.[76] But P Pal likes to identify it as representing Bodhisattva Maitreya and not a Buddha. He, on the basis of literary evidences refers that Maitreya, when preaching in the heaven, appears in crowned and regally dressed.[77] One thing is notable here that Maitreya Bodhisattva is normally shown with the *Amrita kalasa* (Nectar pot) in one of his hands which is conspicuously absent here so the identification with Bodhisattva Matireya does not hold much ground.

(4)   A seated bronze image of the Buddha housed in the Richmond museum is another representative of this period.[78] Here the Buddha is shown seated on the pericarp of a lotus which has three layers of petals moving downwards. The lotus is placed on a simple pedestal. The hands are turned in *Dharmacakrapravartana mudrā*. His head is covered with stylised hairs with prominent *Uṣniṣa*. The elongated eyes of this image is inlaid with silver and pupils in the eyes painted with black lacquer-like paint. The lips of the image are inlaid with copper.

(5)   S.P.S. Museum housed an inscribed Buddha image in broze dated in the eighth century A.D. is the next representative of this period. This crowned Buddha image seated in yogic posture is shown with a typical hand pose which was noticed in some

---

76. *The Way of Buddha*, Delhi, 1956, p.193, fig.12.

77. P. Pal, *Bronzes of Kashmir*, New Delhi, 1975, p.116.

78. B. Malla, *Op. Cit.*, pl.36, p.46.

of the images of early Gandhāra school. The image is very much similar to the posture of *Dharmacakrapravarttana mudrā* but it is technically called *Bodhyangi mudrā*. This typical hand pose is prescribed in the *Sādhanamālā* for Vairocana in which two hands held against the chest with the tips of the thumb and forefinger of each hand united.[79]

(6)   A seated images, housed in the Kashmir Museum is the next representative of this period(pl.15). Here the Buddha is shown seated on a lotus throne and the hands are in *Dhyāna mudrā*. The drapery is foldless which covers both the shoulders. The ears are elongated and the hair is curly with prominent *Uṣniṣa*. The image is rubbed in the front side slightly and is sliced into three pieces which has been arranged in a systematic manner.

(7)   The next representative of this period is a tiny standing bronze image, housed in the Srinagar Museum (Pl.17). The general features and treatment of the garment clearly manifest the Gandhāran influence on the image. Here the Buddha is shown in *Abhaya-mudrā*. The folded drapery covers both the shoulders. The eyes are open, the ears are elongated and the hair is arranged in waves with prominent *Uṣniṣa*. There is *Ūrṇā* between the eyebrows. The overall treatment of the image is not very much impressive.

---

79. B. Bhattacharya, *Indian Buddhist Iconography*, Calcutta, 1968,pp.53-54; P. Pal. *Op. Cit.*, p.120.

CHAPTER - 6

ICONOGRAPHY OF THE BUDDHA IMAGES IN WESTERN INDIAN CAVES

6.1 : *WESTERN INDIA : Historical perspective of the region.*

6.2 : *Buddhism in Western India.*

6.3 : *Western Indian Buddhist Caves and Sculptures.*

6.4 : *Iconography of the Buddha images.*

6.1: WESTERN INDIA : Historical perspective of the region.

Western India is described in the Brāhmanical text *Kāvyamimānsā* of Rājasekhara as *Devasabhāyahparatāh paścatdesaḥ* (the country lying to the west of Devasabhā). The *Bhuvanakoṣa* sections of the *Purāṇas* indicate that Western India was known by the name of Aparānta.[1] Early Buddhist texts contain a scanty account of Aparānta, perhaps due to its location beyond the Majjhimadeśa, the most sacred tract to the Buddhist writers. But the term Majjhimadeśa itself indicates that the five-fold divisions of India might not be unknown to the early Buddhists. R.G. Bhandarkar says that Aparānta was the Northern Konkan whose capital was Sūrpāraka.[2] According to the *Mārkaṇḍeya Purāṇa* Aparānta might be situated in the 'Sindhu-Sauvira' counrtry.[3] Aparānta is also mentioned in the *Mahābhārata*.[4] Aśoka's V Rock Edict mentions the name of Aparānta and it is also mentioned in the Nāsik record of Gautami Balasrī.[5]

---

1. S. Sengupta, "Buddhism in Western India", *The Mahabodhi*, Vol. 63, Calcutta, p.243; B.N. Chaudhury, Buddhist Centres in Ancient India, Calcutta, 1982, p. 169.

2. R.G. Bhandakar, *Early History of Deccan*, p.25.

3. B.N. Chaudhry, *Op. Cit.*, p. 169.

4. *Ibid*.

5. D.C. Sircar, *Select Inscriptions*, Vol. I, Calcutta, 1942, pp.23,196.

Huien-Tsang[6] mentions that Western India comprised Sindh and Western Rajasthan with Kutch and Gujarat and a portion of the adjoining coast on the lower course of the Narmada river - three states; Sindh, Saurastra - Gujarata and Maharastra.

The history of Western India before the rise of the Mauryas is not very much clear due to the paucity of adequate materials. During the period of *Mahājanapadas* Avanti was an important kingdom of Western India.[7] Ancient Avanti comprised modern Malwa and parts of Madhya Pradesh. It was divided into two parts: North and South Avanti. North Avanti had its capital at Ujjain, and Mahismati was the capital of South Avanti.[8] During the days of the Buddha, Pradyota was the king of Avanti and it was an important centre of Buddhism. It is said that during the reign of Pradyota the kingdom of Avanti was in rivalry with the neighbouring kingdoms of Vatsa, Kosala and Magadha. The *Mahāvagga* mentions that he was a cruel ruler. All these born out by his epithets, *Chanda* and *Mahāsena*. Pradyota ruled for twenty three years and he was followed by four kings viz., Palaka, Visākhāyupa, Ajaka and Nandivardhana who ruled for twenty four, fifty, twenty one and twenty years, respectively. The last ruler was defeated by Śiśunāga of the Magadhan empire who also included Avanti in his empire and from this period onwards Avanti became the part and parcel of the Magadhan empire.[9]

The Ceylonese Chronicles mention that Avanti was the residence of the Mauryan king Aśoka when he was the Governor of Avanti.[10] It is said that after the disintegration of the Mauryan Empire, rather after the death of Aśoka, Western India came under the power of Sātavāhanas who were known as Andhra in the *Purāṇas*. Though there is dispute among scholars regarding the dating of the advent of the Sātavāhana rule in Western India but the opinion of H.C.

6. T. Watters, *On Yuan Chwang*, Vol. II, London,1905,pp.25L ff.

7. H.C. Raychaudhury, *Political History of India*, Calcutta, 1950 (reprint), p.144.

8. Bela Lahiri, *Indigenous States of Northern India*, Calcutta, 1974, pp.85-86; D.N. Jha & K.M. Shrimali (Ed.), *Prāchin Bhārata kā Itihāsa*, (Hindi), Delhi, 1986(reprint), p.164.

9. R.C. Majumdar (Ed.), *The Age of Imperial Unity*, Bombay, 1953, p.30.

10. *Dipavaṁsa*, V.15; *Mahāvamsa*, VIII, 1,27.

Raychaudhury[11] that the Sātavāhana rule began in the first century B.C., has been accepted by the majority of scholars. It seems that the early Sātavāhana rulers appeared in Maharastra where most of their early inscriptions have been found. But there are differences of opinion among scholars that whether Sātavāhanas first flourished in Andhra or in Western India mainly due to the lack of adequate evidence. However, it is quite clear that the Sātavāhana empire comprised not only considerable part of Western India, but also the part of South India. The first rulers of this dynasty was Simuka (C.30 B.C.) who ruled for twenty years. He was succeeded by his brother Krisna who ruled for eighteen years and followed by Sātakarni I who also ruled for eighteen years. After that the Sātavāhanas were uprooted by the Śaka ksatrapas from the Western Deccan. The coins and inscriptions of Śaka-Ksatrapa Nahapāna which have been discovered in the surrounding areas of Nasik, show Śakas domain over those areas either in later first century A.D. or in early first century A.D.[12]

The Śaka rulers Nahapāna ruled for six years (119- 125 A.D.) He was defeated and killed by the Sātavāhana king Gautamiputra Sātakarni (C.106-30). He is said to have annexed the southern provinces of the Western Ksatrapas. Gautamiputra was succeeded by his son Vasisthiputra Sri Pulumavi whose inscriptions have been found in Nasik, Karle and Amarāvati.[13] During the reign of Pulumavi there had been a war between Sātavāhanas and the Kardamakas, another dynasty of the Śakas by which Sātavāhanas were defeated and it is also said that a considerable portion of the Sātavāhana territory was captured by the Kardamakas. Pulumavi was succeeded by Śivaśri Sātakarni (159-166 A.D.) who is usually identified with Vasisthiputra Sātakarni, son- in-law of Rudradamana.[14] He was followed by Śivaskanda Sātakarni (C.167-174 A.D.)[15] who twice waged war against Rudradaman. Śivaskanda was succeeded by Yajñaśri Sātakarni (C. 174-203 A.D.) who is considered as the last great Sātavāhana ruler. In

---

11.   H.C.Raychoudhuri, *Op. Cit.*, p. 408.

12.   D.N.Jha & K.M.Shrimali (Ed.), *Op. Cit.*, p. 234.

13.   *Ibid*

14.   R.C.Majumdar(Ed.), *Op. Cit.*, p. 205.

15.   *Ibid*.

the third century A.D. the Sātavāhana power began to decline [16] and many local rulers established their own states. The Ābhiras is said to have snatched Maharastra from the Sātavāhanas.

After the fall of the Sātavāhana empire and before the rise of the Chalukyas, the Vākāṭaka were the most important power who held their sway over parts of the Deccan and sometimes portions of Central India. The founder of this dynasty was Vindhyaśakti who seems to be a feudatory of the later Sātavāhanas of Vidarbha.[17] He is mentioned in the Ajanta inscription of the time of Harisena.[18] Vindhyaśakti was succeeded by his son Mahārāja Hariputra Pravarasena I. In the Purānic accounts he is mentioned as Pravira.[19] He is the only Vākāṭaka ruler who has been called as a *Samrāṭa* (a universal monarch) in some records. He was the great champion of Brāhmanical religion and is credited to hold many Vedic sacrifices viz., *Vājapeya, Aśvamedha* etc. After the death of Pravarasena I in about early first century A.D., the Vākāṭaka empire was divided into two parts. One part of the empire came under his son Sarvasena who made Vatsagulma, in the Akola District, their capital[20] and another part of the empire came under the descendants of his son Gautamiputra who made Nandivardhana (Nagpur) their headquarters. Rudrasena I who succeeded his grandfather, Gautamiputra, is said to have flourished before the victorious advance of Samudragupta in Central India. He was succeeded by his son Prithivisena I and after him Mahārāja Rudrasena II became the king who married Prabhāvatigupta, the daughter of Gupta king Chandragupta II.[21] Rudrasena was succeeded by his minor son, Divākarasena. His wife Prabhāvati ruled as guardian of her son. The Vākāṭaka rule lingered upto the later half of the sixth century A.D. After the Vākāṭakas, the Chalukyas of Badami held their sway over Deccan and parts of South India. They played important role over this region till the middle of the eighth century A.D. when they were overthrown by the

---

16. *Ibid.*, pp. 206-211.

17. *Ibid.*, p. 218.

18. D.C.Sircar, *Op. Cit.*, p. 426.

19. R.C.Majumdar (Ed.), *Op. Cit.*, p. 220.

20. R.C.Majumdar (Ed.), *The Classical Age*, Bombay, 1954, p. 177.

21. R.C.Majumdar(Ed.), *Op. Cit.*, pp. 179- 180.

Rāṣṭrakuṭas. Pulakeśin II is considered as the greatest ruler of the Chālukya dynasty. He is known to us from his eulogy written by the court poet Ravikirti in the Aihole inscription.[22] He is said to have defeated the army of Harṣavardhana of Kanauj on the Narmadā and checked their advance towards the Deccan.[23]

6.2   Buddism in Western India.

The earliest evidence of Buddhism in western India is not beyond controversy. However, it seems that Buddhism might have spread in western India during the life time of the Buddha. But there is no evidence that the Buddha himself visited Western India. The westernmost country visited by the Buddha himself was Verañjā a place near modern Mathura where he spent one of the rainy seasons (*Vassāvāsa*). It is said that when Pradyota, king of Avanti heard about the presence of the Buddha at Verañjā, he sent his head priest Mahākaccāna to meet the Buddha and to invite him to visit Avanti. But before the arrival of Kaccāna at Verañjā, the Buddha had already left for Vārānasi; then Kaccāna also left for Vārānasi and there the meeting was held between them. Kaccāna is said to have become very much influenced by the Buddha and subsequently he became a convert.[24] After his initiation the Buddha also admired the scholarship of Kaccāna to the extent that when the latter requested Him to visit his place, the Buddha replied that there is no need for Him to undertake the journey as you (Kaccāna) yourself are capable of propagating the Dhamma.[25] Subsequently on his return Kaccāna propagated the Dhamma in western India.[26] In this way Buddhism spread to Western India for the first time. The traces of Buddhism in the pre-Aśokan period in Western India is further corroborated by the evidence that when the dispute arose in the Buddhist saṅgha which caused to be held the second Buddhist council in which the monks of Kausambi, Paṭheyya and Avanti were consulted in resolving

22.   D.C.Sircar, *Op. Cit.*, Vol.II, Delhi, 1983, pp.443-448.

23.   R.S.Tripathi, *History of Kanauj*, Varanasi, 1927, pp. 110-112.

24.   S.Sengupta, *Op. Cit.*, p. 243; B.N.Chaudhury, *Op. Cit.*, p. 170; R.C.Sharma, *Buddhist Art of Mathura*, Delhi, 1984, pp. 40-41.

25.   S.Sengupta, *Op. Cit.*

26.   R.C.Sharma, *Op. Cit.*, p. 41.

the problem. These three places are considered as part of Western India.[27]

But Buddhism is said to have been systematically preached in western India during the reign of Aśoka. It is well known that after the third Buddhist council held at Pāṭaliputra under the patronage of Aśoka, Buddhist emissaries were sent to different parts of the country and abroad to spread the Dhamma and that Dhammarakṣita was deputed to Aparānta. It is notable here that Dhammarakṣita is recorded to be a Yonaka or an Ionian Greek. This is probably the earliest reference to a foreigner embracing Buddhism as his religion. Dharmmarakṣita is said to have converted a large number of persons in Western India.[28] The traces of Buddhism in western India in the Aśokan period and onwards is further corroborated by the discovery of Aśokan edicts at Girnar and the existence of the rock-cut cave temples belonging to the second and the first centuries B.C., scattered all over the western India. These caves are the living witness of the spread of Buddhism in this area.[29] But the condition of Buddhism in this region is not precisely known for a long period from the time of the kṣatrapas upto the rise of the Guptas, due to lack of adequate evidences. However, the inscriptions of the kings of Valabhi, accounts of the Chinese travellers and the inscriptions of the caves of Kanheri etc. tell us about the condition of Buddhism during the period beginning from the fifth century to the later half of the eighth century A.D. The inscriptions of Valabhi testify that the Queen Duddā, cousin of king Dhruvasena I (519-549 A.D.), erected a *Vihara* near Valabhi known as Duddā *Vihāra* . The Duddā *Vihāra* was Very large and was often mentioned as *vihāra maṇḍala*, may be a cluster of *vihāras*. The Maitraka rulers of Valabhi were undoubtedly the great champions of Buddhism Some of their inscriptions dated between fifth and seventh century A.D. inform us about the flourishing condition of Buddhism at Kanheri.[30]

---

27.   P.V.Bapat (Ed.), *2500Years of Buddhism*, Delhi, 1976(reprint), pp. 37-38; K.A.Nilakanta Sastri, *Age of the Nandas and Mauryas*, Delhi, 1967, pp. 299-300

28.   *The Mahāvaṁsa*, xii.4.

29.   S.Sengupta, *Op. Cit.*, p. 244; B.N.Chaudhury, *Op. Cit.*, p. 170.

30.   *ASWI*, Vol. V, and vi.

The reign of the Sātavāhanas proved to be a land mark in the history of Buddhism in Western India. During this period world famous caves of Ajanta were excavated, but they did not propsper much and was deserted for a period of about four centuries. It was later revived by the Mahāyānists from the fifth century A.D. then it turned to be an important centre of Buddhism.[31]

The Vākāṭaka rulers, who also ruled over Western India, were great patrons of Buddhism. During their period Ajanta became a much popular centre of Buddhism. A fragmentary inscription from the cave No. 16 at Ajanta records that the cave was excavated by the order of Vīradeva, a devout Buddhist and minister of the Vākāṭaka king Hariṣeṇa.[32] An inscription found in cave No. 17, records that Achitya, a minister of Rabisamba, a feudatory to the Vākāṭaka king Hariṣeṇa, caused to excavate the 'monolithic, gem-like hall' with *Caitya*, a reservoir with cool refreshing water and a Gandhakuṭi.[33] Another important inscription from cave No. 26 tells us that it was the monk Buddhabhadra who was responsible for the excavation of the concerned cave by providing funds for the work. From the inscription it is quite clear that Buddhabhadra was a much esteemed recluse and might have been an abbot of a great institution and *Sthavira* Achala was the former builder of the *Vihāra*. A few inscriptions tell us about the gifts of ministers, noblemen, lay devotees and monks etc.[34]

Ellora was another great centre of Buddhism. The cluster of caves at Ellora are divided among three religions : Buddhism, Brāhmanism and Jainism. The Buddhist caves of Ellora are contemporary with the later phase of those at Ajanta and they reflect Mahāyānic influence.

The Chinese traveller says that when Buddhism was in decaying condition almost all over India, it was still very much prevalent in Western India. Huien-Tsang found monasteries and followers of the

---

31. B.N. Chaudhury, *Op. Cit.*, p.171.
32. S. Sengupta, *Buddhism in the Classical Age*, Delhi, 1985, p. 39; *ASWI*, Vol.IV, pp. 130ff.
33. *ASWI*, iv, pp.130ff.
34. *Ibid.*, pp. 60,124 ff.

Hīnayāna, Sammitiya and the Mahāyāna scholars throughout Western India.[35]

From the eighth century onwards Buddism began to decline in Western India when the Muslims occupied Sindh and adjoining areas, however, still there were certain places in this region where later traces of Buddhism have been perceived.

6.3 : Western Indian Buddhist caves and sculptures:

The earliest caves in India probably are those which are located in eastern India, the small series near Gayā consisting of the Barābar and Rajgir groups dating to the time of Aśoka and these are closely followed by the caves of Orissa - Udayagiri and Khandagiri.[36] The caves of Kathiawar region - Junagarh, Talaja and Sana are next in order and these are followed by the caves of Bhaja, Kondane, Bedsa and Karle.

The caves of Western India have broadly been divided into two sections: 1. Hīnayāna and 2. Mahāyāna. The caves of Hīnayānic phase containing the *Stūpa* in the sanctum sanctorum whereas the Mahāyānic caves contain the images of the Buddha etc. Here we have taken only a few caves in our account which contain the images of the Buddha viz., Karle, Nasik, Kanheri, Bagh, Ellora and Ajanta.

KARLE : Karle is situated in the Borghata hills between Bombay and Poona. The most magnificent and well-preserved Buddhist caves have been discovered here. They usually belong to the late Hīnayāna period i.e. first or second century A.D. But the influence of the Mahāyāna is also traceable here in later additions.[37] They consist of a large *Caitya* and several *Vihāras*.[38] There are many inscriptions in about the caves which usually refer to the names of donors, devotees etc. On stylistic and palaeographic considerations, these caves have been placed to the period of early second century A.D.[39] Fergussan

---

35. T. Watters, *On Yuan Chwang*, Vol.II,London,1905, pp.241-248.

36. J.C. Harle, *The Art and Architecture of the Indian Subcontinent*, Penguin Books, 1986, pp.43-45;Vidya Dehejia, *Early Buddhist Rock Temples...*, London, 1972, p.71; S.R. Wauchop, *The Buddhist cave Temples of India*, New Delhi, 1981,p.18.

37. S.R. Wauchop, *Op. Cit.*, p.40.

38. *Ibid*.

39. D. Mitra, *Buddhist Monuments*, Calcutta,1971, p.154.

says that "it was excavated at a time when the style was at its greatest purity. In it all the architectural defects of the previous examples are removed; the pillars of the nave are quite perpendicular. The original screen is superseded by one in stone ornamented with sculpture, its first appearance apparently in such a position, and the architectural style had reached a position that was never afterwards surpassed."[40] At the entrance of the cave a pillar stands which is surmounted by four lions. On the right hand side, there is a Śiva Temple. It is said that previously there was a second pillar on the spot of the Śiva Temple which was surmounted by a wheel.[41] The outer porch of the building is wider in comparision to the body of the building and is closed in front by an outer screen, composed of two stout octagonal pillars, without either base or capital, supporting a plain mass of rock, but once it was ornamented by a wooden gallery, forming the principal ornament of the facade. The facade of the hall at the back are embellished with carvings, among which stand out six stately large couples of robust vitality and majestic bearing. In the side walls are introduced the foreparts of three elephants, standing above a railing motif in the attitude of supporting on their shoulders the superstructure, consisting of reliefs of several storeys of barrel vaulted structures with *Caitya* - windows and railings. The reliefs of the Buddha, in some cases at the expense of the earlier reliefs, are palimpsests of about the sixth century A.D. which have marred considerably the balanced effect of the original scheme.[42] Above this, the *Caitya* arch and 'rail pattern, are repeated again and again to the top. On the front wall of the cave both the 'rail' at the bottom and that on a level with the heads of the doors have been cut away in later times to make room for images of the Buddha and his attendants, Padmapāṇi, etc. In the middle of the space between the central and right hand door is inserted of a sculpture which may be of a later date; the Buddha is there attended by Padmapāṇi and perhaps Mañjuśri seated on the *Siṁghāsana* with his feet on the lotus over a conventionalised wheel, supported by two deers and under the wheel

---

40. James Furgussion and Burgess, *Cave Temples of India*, London, 1880, p.232.
41. S.R. Wauchop, *Op. Cit.*, p.41.
42. D. Mitra, *Op. Cit.*, p.155.

is a supporting tier held by Nāga figures while over the Buddha's head two Vidyādharas hold a tiara.[43]

There are three doors, one leading to the centre and one to each of the side aisles and over the gallery the whole of the hall is open as in the *Caitya* halls forming one great window, through which all the light is admitted. The interior arrangement of the great *Caitya* follows the common plan of this class, but the original wooden inverted 'U' and half 'U' shaped beams and longitudinal purlins of the framework, the vault of the nave and apse are of obvious interest.[44] There are fifteen pillars on each side separating the nave from the aisles. Their bases consist of the *ghaṭa-base* (water pot), the shaft is octagonal representing the Saṅgha while the capital is the seed of the sacred lotus supporting as in the ancient Vedic altar, in this case, two elephants, guardians of the southern quarter bearing two figures, generally a man and a woman, but sometimes two females very much better executed. Above the sculptures on the capitals spring the roof, semi-circular in general section, but somehow, stilted on the sides, so as to make its height greater than the semi-diameter.[45] The *Stūpa* or the apse has a drum in two terraces, both crowned by a railing. The bottom side of the wooden umbrella, presumably coeval with the excavation, is carved minutely with delicate patterns including a lotus.

The caves of Karle are not numbered in any particular sequence. The cave No. 2 is a three storyed *Vihāra* just to the left of the above mentiioned *Caitya*. The top storey has a verandah with four pillars, with slightly ornamented capitals. On the left side in the top storey is a raised platform in front of five cells with slots for a beam along the front and apparently each cell could also be isolated. The cave No. 4 is situated to the south of the *Caitya* and from an inscription it is learnt that it was given by Haraphana, a Persian, during the reign of the Sātavahāna king Gautamiputra Pulumavi.[46]

NASIK : Nasik is situated about 75 miles to the North-west of Bombay. The Buddhist caves here known as Paṇḍuleṇa, are very

---

43. *Ibid*.
44. *Ibid*.
45. *Ibid*., pp. 155-156.
46. B.N. Chaudhury, *Op. Cit.*, p.179.

much famous. There are twenty three caves in number and it is said that it was excavated by the Bhadrayānika sect of Hīnayāna tradition.[47] The earliest one is probably a *Caitya* cave which belongs to the Christian era and these caves were in existence till the sixth or seventh century A.D.[48] Both the Sātavāhanas and the kṣatrapas were the patrons of this establishment. The excavation of the two largest monasteries is credited to them and the remaining caves were the gifts of the common people, monks etc.

The cave No. 17 is placed to the period of first century B.C. - A.D.[49] The hall of cave measures 23 feet wide by 32 feet deep and has a back aisle screened off by two columns. The verandah is somewhat peculiar, it is reached by half a dozen steps in front between the two central octagonal pillars with very short shafts, and large bases and capitals. On the wall of the back aisle there is a standing image of the Buddha. On the right side there are four cells without branches. There is an inscription which records that the cave was the work of some Indrāgnidatta of Dattamātri. The cave No. 20 is a larger *Vihāra* and it has eight cells on each side, while at the back are two cells to the left of the ante-chamber and one to the right, with one more on each side of the ante-chamber and entered from it. There is a colossal image of the Buddha in the preaching *mudrā*. Here the Buddha is seated in *Bhadrāsana* attended by Padmapāṇi and Vajrapāṇi.[50] An inscription of the seventh year of Sātavāhana king Yajña Sātakarṇi states that the cave's initial excavation was first started by an ascetic but completed by the wife of a mahā-Senāpati.[51]

The cave No. 23 is considered as another interesting cave, but the whole facade of the cave is destroyed.[52] All the shrines as well as many compartments on the walls are filled with sculptures of the Buddha attended by Padmapāṇi and Vajrapāṇi. In the inside of the cave there are also the images of the Buddha in different *mudrās*. The

---

47. J.C. Harle, *Op. Cit.*, pp.52-55; S.R. Wauchop, *Op. Cit.*, p. 65; B.N. Chaudhury, *Op. Cit.*, p.181.

48. D. Mitra, *Op. Cit.*, p. 168.

49. S.R. Wauchop, *Op. Cit.*, pp.68-69; D. Mitra, *Op. Cit.*, pp. 168-169.

50. D. Mitra, *Op. Cit.*, p.169.

51. *Ibid.*

52. S.R. Wauchop, *Op. Cit.*, p.71.

notable feature of the Buddha image of *Vyakhyāna mudrā* is that it has moustaches.[53]

KANHERI : Kanheri is situated about twenty miles north of Bombay. It was a great centre of Buddhism for several centuries. There are more than hundred caves at this place which was also a large monastic establishment. The architecture is dated from the beginning of the Christian era and as late as the eleventh century A.D.[54]

Generally the caves are small and consist of a court with a recess in one of its side walls over a cistern, a raised pillared verandah approached from the court by a flight of steps with a moonstone at the base and a cell or an astylar hall with or without windows, the latter often graded in the earlier examples. There are a few cells by its sides in the hall of the larger caves. Edging the pillars of the verandah is often a parapet relieved with a railing resting on beams and rafters; the facade of the plinth below is panelled with pilasters in reliefs.[55]

The caves No. 41,67,89 and 90 etc. have on their walls a dazling profusion of reliefs, mostly the Buddha. Here the facial expression of the images of the Buddha reflect the transcendental bliss. The Buddha is shown both in standing and seated postures. The right hand of the standing Buddha images are turned in the *Varadamudrā* whereas the seated images are in the *Dharmacakrapravartanamudrā*. Here too the Buddha is seated in *Pralambapādaāsana*. The Buddha is flanked by two Bodhisattvas with female deities, probably their Śaktis.[56]

The other deities which have been depicted on the caves are Śakra, the god of gods, Panchaśikha etc. (in the cave No.90), Avalokiteśvara in the company of female deities (in the cave No.2,41 and 90), and some other Buddhist deities. One of the reliefs of the cave No. 67 depicts the *Dipankara Jātaka*.[57]

---

53. *Ibid*
54. D. Mitra, *Op. Cit.*, pp. 164-165.
55. *Ibid.*, p.165.
56. *Ibid*.
57. *Ibid.*, pp. 165-166.

It is very difficult to fix the date of each cave, but it is evident that they continued to be in use at least till the eleventh century A.D.[58] It is notable here that a modern Japanese inscription of a Buddhist pilgrim of the Nichiren sect is engraved on the walls of the cave No. 66 which testifies the continued occupation of these caves even in modern times.[59]

BAGH : The caves of Bagh are situated in the south of Mālwā about 25 miles South-west of Dhar District and 90 miles to the West-south-west of the Mhow railway station.[60] The number of caves situated at Bagh are nine in number and extended over a frontage of about 750 yards.[61]

Architecturally, these caves are contemporary to the caves of the Deccan, but they also have some singular traits. The cave No.2 which is popularly known as the 'Pāṇḍavoṅkī Gumphā' is one of the best preserved in the group. It is a square *Vihāra* with cells on three sides and a *Stūpa* inside a shrine in the back, the total measurements from front to back being rather more than 150 feet. Its facade was originally relieved with *Caitya*-windows having insets of tiger-heads and lotuses. The ante-chamber has two 12-sided pillars in front. The walls of this room are adorned with sculptures. On each end there is a standing image of the Buddha in *Varadamudrā* flanked by two attendants.[62]

Cave No. 3 which is locally known as the *Hathikhana* is also a *Vihāra*. In comparison to previous one its cells are of a more eloborate design. Much of the forefront of the cave has fallen, but it is obvious that there must have been a row of chambers on the South-west side corresponding with those on the North-east.[63] The cave originally consisted probably of its distinct halls, an outer one supported on eight octagonal pillars, row of four cells on the right side of the hall, a pillard hall, probably a later addition, at the back side, and at the

---

58. *Ibid.*, p. 166.
59. P.V. Bapat (Ed.), *2500 Years of Buddhism*, Delhi, 1976 (reprint), pp. 291-292.
60. D.Mitra, *Op. Cit.*, p.99.
61. S.R. Wauchop, *Op. Cit.*, pp.83-84.
62. D. Mitra, *Op. Cit.*, pp.99-100; S.R. Wauchop, *Op. Cit.*, pp.83-84.
63. D. Mitra, *Op. Cit.*, p.100.

left side a complex of cells comprising a pillard vestibule leading to a rear chamber, with painted figures of the Buddha and kneeling devotees on its walls, and four cells, two each on either side of the chamber but separated by a passage. On the left side of the court, a similar complex with painted figures of the Buddha on the walls of the central chamber is present.

The remaining caves are not concerned to our present study.

ELLORA : The famous caves of Ellora which are spread over a distance of about 2 kilometres, from south to north, on a sloping hillside, known after the adjacent village of Ellora, is situated about 16 miles from Aurangabad.[64] The caves of Ellora is divided into three groups. Buddhist - 12, Jain - 5, and Brāhmanical - 17 which are located here side by side. The twelve Buddhist caves are all *Vihāras* except the *Viśvakarmā* which is a *Caitya* hall enshrining a *Stūpa* carved with a colossal image of the Buddha flanked by two attendants. They are contemporary to the later phase of Ajanta.

Cave No. 2 is larger and its individualistic treatment is noteworthy. It consists of a verandah, the front of which has been carved in compartments with fat *gaṇas* or dwarf figures, often in grotesque attitudes. The cave contains lateral galleries filled with the images of the seated Buddha in *Pralambapādaāsana* and in *Vyākhyāna mudrā*. The right porch has a relief of the Great miracle of Śrāvasti. The roof is supported by twelve massive pillars, arranged in a square, have each a high square base, the capitals, which may be termed as the cushion capital. The doors of both the hall and the shrine are guarded by large images of Bodhisattvas. On the front wall of the hall there is a profusion of the reliefs which depict the Buddha and other divinities of Buddhist pantheon.[65]

Cave No. 11, which is locally known as *Do Thāl* (two- storied), is actually a three-storied monastery. The ground floor of the cave was excavated in 1876.[66] In contrast to the plain facade, the inner halls of this cave are richly decorated with the figures of the Buddha, the Bodhisattva and some Vajrayāna deities. In the second chamber

---

64. Ibid., *Op. Cit.*, p.181.

65. Ibid., *Op. Cit.*, pp.181-182; S.R. Wauchop, *Op. Cit.*, pp. 87ff.

66. J. Burgess, *A Guide to Elura cave Temples*, Hyderabad, 1926, p.16.

of the first floor, a colassal image of the Buddha in *Bhumisparśamudrā* is flanked by two Bodhisattvas.[67]

Cave No. 12 which is locally known as *Teen Thāl* (three storied) is probably one of the latest caves belonging to the Vajrayāna pantheon and is noteworthy for its abundant sculptured peculiarities from the iconographic point of view, carved approximately in the first half of the eighth century A.D. In this cave Avalokiteśvara appears as a *Dvārapāla*. The shrine contains a colossal image of the Buddha. The divine Bodhisattvas, eight in number, are carved on the side walls. Here we notice a change in the style of the sculptures, e.g., elongated faces etc. Each storey is of considerable size and designed with great accuracy. They each contain the images of the Buddha with his usual attendants. Each storey is also approached by a well-made staircase from the storey below. They contain cells in the walls.[68] It is said that *Teen Thāl* was a teaching place.

AJANTA : The rock-cut Buddhist caves of Ajanta which are situated about 60 miles to the North-east of Aurangabad, have a most significant place in Indian art. The artistic activity in the caves is said to have started at least as early as the second century B.C.[69] and continued as cultural centres upto the seventh century A.D. The caves, including the two inaccessible ones, are thirty-one in number and are considered as a superb artistic unity of architecture, pictorial and sculptural form. The later caves of Ajanta, carved from the fifth century A.D. onwards, are modified to suit the objectives of the Mahāyāna faith, prior to the advent of the Vajrayāna influence.[70] It is in these later caves we are confronted with the introduction of Mahāyānic deities including the images of the Buddha.

Cave No. 4, although unfinished, is the largest *Vihāra* at Ajanta and can be assigned to the first half of the sixth century A.D. To the right of the elaborately decorated doorway, in a rectangular panel between two pillars, the litany is carved with Avalokiteśvara in the centre of the eight great perils, represented by four scenes on either

---

67. D. Mitra, *Op. Cit.*, p. 186.
68. *Ibid.*, pp. 186-188.
69. D. Mitra, *Ajanta, ASI*, New Delhi, 1983(reprint), p.6.
70. S.L.Weiner, *Ajanta Its Place in Buddhist Art*, California, 1977, p.8.

side. Avalokiteśvara in the *Jaṭāmukuṭa* of which an image of the Buddha, seated in the *Vyākhyānamudrā* is seen.[71]

Inner shrine of the same cave contains a colossal image of the Buddha flanked by two attendants, Avalokiteśvara on the right side and Vajrapāṇi on the left. The trinity is attributed to the mid-sixth century A.D. on the basis of a votive inscription depicted on the pedestal of the Buddha image.[72]

Cave No. 10 is the earliest *Caitya-griha*, dating, according to the inscription from the early second century B.C. later, in the sixth century A.D. Mahāyāna Buddhism reveals its traces on the upper facade. To the right side, an outstanding Buddha trinity in a rectangular niche can be marked. The Buddha in *Dharmacakra mudrā* is flanked by two attendants.[73]

Cave No. 17 was excavated during the reign of Harisena, a feudatory prince of the Vākāṭaka king as stated in an inscrition on the wall of the Verandah.[74] In the shrine, alluded to as *Muni-rāja-Caitya* i.e. a *Caitya* of the ascetics.[75] Here we find a colossal image of the Buddha in *Dharmacakramudrā*, flanked by two standing *cāmara-*bearing Bodhisattvas. On the pedestal, the wheel and the deer symbol is depicted and there are two standing images of friars in front.

Another Buddha trinity is carved on the outer wall of this cave.

The far-reaching fame of Ajanta is for its fresco painting which were found somewhere mutilated, but later on it were restored. The painting illustrates in addition to decorative designs, scenes from the life of the Buddha and the *Jātaka* and as also scenes from secular lives and are living specimens of Buddhist master artists bearing the best expressions of grace and loveliness which engaged the attention of scholars throughout the world.

71. *Ibid.*, Chapter one.
72. *Ibid.*, p. 101.
73. D.Mitra, *Op. Cit.*, pp. 42-45.
74. S.L.Weiner, *Op. Cit.*, p. 8.
75. *Ibid.*, p.113.

6.4   Iconography of the Buddha images:

Though much attention has not been given on the merits of the rock sculptures, may be due in part, to its being woven with rock architecture, even the Buddhist sculptures of these caves have not been more systematically categorized into separate school, but the images of the Buddha of Western Indian caves acquired a magnificent place in the history of Indian art.[76]

The characteristic features of the images of the Buddha of the Western Indian caves:

1. The images of the Buddha are represented mainly in three *mudrā*. The most usual attitude in which the Master is represented as seated on a throne, the corners of which are upheld by two lions and consequently known as the *Singhāsana'* or 'lion-seat', with his feet on a lotus blossom and his hands in front of the chest holding the little finger of the left hand between the thumb and the forefinger of the right hand. This is known as *Dharmacakrapravartana mudrā*. The next common *mudrā is Dhyāna-mudrā* in which the Buddha is seated in cross-legged attitude and the palms turned upwards and the eyes are half closed. In the third *mudrā* the Buddha is shown seated in *Vajrāsana* and his left hand lies on the upturned soles of the feet and the right hand resting over the knee, points to the earth.

2. The Buddha is also sculptured on the walls, where he is shown standing with the alms bowl of the monk.

3. Here the Buddha is sometimes seated in *Pralambapādaāsana* in which the Master is shown seated in a chair-like thing and his both legs hang down vertically from the seat. But this type of seat or *āsana* was not only shown here, it also can be seen both in Mathura and Sārnāth. It is also well known that the fragmentary image of the Kuṣāṇa king Vima Kadaphises which is housed in the Mathura Museum, in which the king is seated

---

76. M.K.Dhavalikar, "Problems of Rock Art," in D.Handa(Ed.), *Recent Studies in Indology*, Vol.II.Delhi, 1989, pp. 141-142.

      in the *Pralambapādaāsana*. This is also known as *Bhadrāsana* and 'European style'.

4.   Another *mudrā* is lying in *Mahāparinirvāṇa* in which the Buddha is shown that He is resting on His right side with head to the north. This *mudrā* usually refers to the death or *Mahāparinirvāṇa* of the Buddha.

5.   The front of the throne, on which the Buddha is shown seated, is usually sculptured with wheel, turned edgewise to the spectator with a deer couchant on each side which most probably represent the deer park of Sārnāth. Sometimes behind the deer there are a number of kneeling worshippers on each side.

6.   The standing images which are usually in relief are mainly in *Varadamudrā*.

7.   The seated images of the Buddha are occasionally flanked by Bodhisattvas.

8.   The body of the seated images are rather heavy and sturdy particularly the upper portion whereas the body of standing images are slim, delicate and flexible. The head is oval, and the facial expression shows serenity and bliss which may reflect the Sārnāth influence.

9.   The drapery is foldless which clearly reflect the Sārnāth influence.

10.   The halo is circular and plain. The hair is arranged in curly manner.

11.   Sometimes the Buddha images are depicted on the *Jaṭāmukuṭa* as in the cave No. 2 of Bagh.[77]

Some of the specimens which bear above mentioned features are as follows:

KARLE : 1. On the middle panel of the left part of the facade of the *Caitya* hall, we find a seated image of the Buddha. Here the Buddha is shown seated on a throne and his hands are turned in *Dharmacakrapravartana mudrā*. On the pedestal a wheel and deer symbol

---

77.   D.L.Snellgrove(Ed.), *The Image of the Buddha...* pp. 108-110; S.R.Wauchop, *Op. Cit.*, pp. 14-15.

is depicted which probably refer to the event of the first sermon of the Master. The Buddha is flanked by two Bodhisattvas.[78]

2. In another relief in cave No. 2 we find a seated image of the Buddha in the *Vyākhyāna mudrā* and He is flanked by two Bodhisattvas, among which on the left *Cāmara(or* fly-whisk)-bearing is Padmapāṇi, is represented in standing pose, while his Śakti, probably Tārā, with the lotus-stalk in her left hand stands by his side.[79]

NASIK : 1. On the left side of the east wall of cave No. 2 there is a seated image of the Buddha in *Pralambapādaāsana*. The hands of the Buddha are turned in the *Vyākhyāna mudrā* . On the left side of the Buddha there is an image of the *Cāmara* bearing Padmapāṇi.

2. Another representative is a preaching image of the Buddha in the cave No. 23. On the left side of the image, there is a standing image of Avalokiteśvara who holds a *Cāmara* in the right hand and lotus-stalk in the left. Here the image is expressive.[80]

KANHERI : 1. Cave No. 11 which is known as the Darbāra or Mahārāja cave contains a large number of images of the Buddha. On the north wall of the shrine appears a preaching image of the Buddha in *Pralambapādaāsana*. The Buddha is flanked by two Bodhisattvas. On the left side of the Buddha appears a *Cāmara-* bearing Padmapāṇi while another side a *Cāmara* bearing Bodhisattva is represented.[81]

2. Cave No. 41 contains a seated image of the Buddha in *Ardhaparyaṅka mudrā*. Here also the Budddha is flanked by two Bodhisattvas. Here we find an image of four-armed eleven-headed Avalokiteśvara. It is said that this is only known relief of this form in India, however, the cult of this particular form was popular in China in the seventh and eighth century A.D.[82]

BAGH : 1. In cave No. 2 of the Bagh caves, the side walls of the antechamber, each of which have a bold-relief of the standing Buddha, flanked by two attendants. Here the Buddha is shown

---

78.   D.Mitra, *Op.Cit.*, p. 156; S.R.Wauchop, *Op.Cit.*, p. 42.

79.   S.R.Wauchop, *Op. Cit.*

80.   Ibid., pp. 66,77; D.Mitra, *Op. Cit.*, p. 171.

81.   S.R.Wauchop, *Op.Cit.*, pp. 75-76.

82.   D.Mitra, *Op. Cit.*, p. 166.

standing in *Tribhaṅga* pose on a lotus seat (Pl.21). The drapery, on which there are incised lines, covers the left shoulder only and the lower garment reaches below the knee. The right hand makes the gesture of *Varadamudrā* while the left hand is held akimbo on which the hem of the garment rests. The ears are elongated and the eyes are half closed and the hair is arranged in curly manner with prominent *Uṣnīṣa*.

2.   There is another image of the Buddha in the same cave. Here the Buddha is shown in seated posture and his right hand is turned in the *Abhaya mudrā*, Interestingly here the Buddha is depicted in the *Jaṭāmukuṭa* of the Bodhisattva, placed in the arched niches on either side of the doorway as Dvārapālas.

ELLORA : 1. Cave No. 2 contains a colossal image of the Buddha. The Buddha is seated here on a lion throne (Siṅghāsana) and his hands are turned in the *Dharmacakra mudrā*. He is flanked by two *Camara-* bearing Bodhisattvas. The rocky surface behind the Buddha is decorated with a few animal motifs while the top shows the *Gandharvas*. On each wallside again a colossal Buddha figure appears but this time he is standing in *Dvibhaṅga mudrā*.[83]

2.   Cave No. 10 housed a gigantic figure of the Buddha (Pl.20) On the *Stūpa* a huge image of the Buddha is depicted in *Pralambapādaāsana*, flanked by Avalokiteśvara and Maitreya under the *Bodhi*-tree with a pair of *gandharvas* on each side. The *Stūpa* is surmounted by a capital but no umbrella is depicted. Here the Buddha is not as condensed in height as in cave No. 1 at Ajanta, but nevertheless is quite as mute and insensitive, an effect caused by the hard treatment of the plastic surface and by the heavy, round, log-like legs and arms. Despite its seemingly meditative or introspective attitude, the image fails to articulate the idea of what the Buddha was supposed to be.[84]

3.   On each side of the rectangular hall in the cave No. 2 we find a row of the seven images of the Buddha, arranged on a

---

83. S.R.Wauchop, *Op. Cit.* pp. 88-89.
84. D.L.Snellgrove, *Op. Cit.*, p. 109. pl.84.

platform. On one side, the figures are seated in the *Dhyānamudrā* and on the other side, they are seated in cross-legged manner and making the gesture of preaching(*Vitarka-mudrā*). At the farthest end of the gallery, on a raised platform, there is the image of a Buddha seated on a high lion pedestal in *Yogāsana* and showing the gesture of meditation. The body and limbs are full, round, warm and vibrant.[85] Overall the figures are well impressive, may be considered among the finest images of the Buddha.

AJANTA : 1 Cave No. 10 contains a seated image of the Buddha in *Dharmcakramudrā*, flanked by two Bodhisattvas.

2.   In the cave No. 19 the Buddha is shown seated in *Pralambapādaāsana* and his hands are turned in the *Vitarka-mudrā*, Here the Buddha is surrounded by other small Buddha figures. The body of the Buddha figure has heavy roundness but it also has the other classical features like robe, hair etc.[86] On the facade of the same cave there are other standing figures of the Buddha which is heavy and plump and is shown in walking position and meeting a child.[87]

3.   At the apse of a *Stūpa* in a *Caitya* hall of cave No.26 there is a prominent figure of the Buddha seated in *Pralambapādaāsana*. The figure is carved against the elongated and embellished drum of the *Stūpa* The right hand of the image is missing. The figure is here to be understood as a manifestation of the instrinsically invisible and unimaginable Buddha in His true essence.[88]

4.   In cave No. 2 at Ajanta there is another seated image of the Buddha who is shown in *Pralambapādaāsana* (Pl.18). The hands are in the preaching pose. The drapery is foldless which covers both the shoulders. The ears are elongated, the eyes are open and the hair is arranged in curly manner. The round halo is decorated and other portion of the back piece is engraved with

---

85.   *Ibid.*, p. 109.
86.   *Ibid.* pl. 72
87.   *Ibid.*, pl. 73.
88.   *Ibid.*, pl. 15.

various figures. The pedestal is supported by the lions facing front.

5.   Cave No. 19 of Ajanta possessed a number of standing Buddha images in relief, features of one of them is described here (pl.19). The Buddha is shown here in *Varadamudrā;* the drapery is foldless and covers the left shoulder only, the ears are elongated the hair is arranged in curly manner with *Uṣṇiṣa* and the contemplative eyes are half-closed. On the left corner of the Buddha seat two seated female devotees are depicted in *Añjalimudrā*. The facial expression manifests the serenity and calmness.

CHAPTER - 7

ICONOGRAPHY OF THE BUDDHA IMAGES IN ANDHRA SCHOOL OF ART

7.1 : *ANDHRADEŚA : Historical perspective of the region.*

7.2 : *Buddhism in Andhradeśa.*

7.3 : *Andhra school of Buddhist art.*

7.4 : *Iconography of the Buddha images.*

7.1 : ANDHRADEŚA : Historical perspective of the region.

The history of Andhradeśa and the adjoining portion of South India is not very much clear before the rise of the Mauryas mainly due to lack of adquate material, however, it is believed that before the Mauryas, South India was under the power of the Nandas[1] who ruled from Magadha till when Chandragupta Maurya dethroned Dhanananada, the last Nanda ruler, and occupied the Kingdom of Magadha. The Hāthigumphā inscription of Khāravela mentions the name of one of the Nanda kings in connection with the construction of a canal in Kaliṅga,[2] which was considered as part of South India. So on the basis of this epigraphical source it is said that Kaliṅga was a part of Nanda empire. But B.M. Barua is of opinion that Nanda had not conquered any part of Kaliṅga on the ground that the province "had remained unconquered till the 7th year of Aśoka reign."[3]

During the rule of the Mauryas, particularly during the Aśokan period, some parts of South India came under their domain and the remaining portions were divided into three kingdoms :

1. K.A. Nilakanta Shashtri, *Dakshina Bhārata kā Itihāsa* (Hindi), Patna, 1986, p. 70.
2. D.C. Sircar, *Select Inscriptions*, Vol.I, Calcutta, 1942, p.211.
3. B.M. Barua, "Hāthigumphā Inscriptions of Khāravela", *IHQ*, Vol. XIV, 1938, p.276n.

1. The Chola, 2. The Chera, and 3. The Pāṇḍya.

1. *The Cholas* : The Cholas were one of the oldest dynasties of South India. The Chola country comprised the lower Kāveri valley, the coastal plain between two rivers - both bearing the name vellar; the north vellar entering the sea near Porto Novo, and the smaller southern stream passing through Pudukkottai territory. The Chola kingdom thus roughly corresponds to modern Tanjor and Tiruchirāpalli districts; its inland capital was Uraiyur, and Puhār or Kāveripattinam which was founded by the Chola king Kārikāla, the most distinguished among the Saṅgam Cholas.[4]

Kārikāla is said to have been a very competent ruler and a great warrior. He defeated the Chera king Perunjeral. Kārikāla maintained a powerful navy which he used to conquer Srilaṅka, from where he brought a large number of prisoners of war whom he used for building a huge durable embankment to tame the Kāveri river. He made Kāveripattinam an important port and an alternative capital of the Chola Kingdom. After the death of Kārikāla his state was steeped in utter. confusion as a result of domestic strife in the Chola family. The last great Chola ruler was Nedunjelian who successfully fought against the Pāṇḍyas and Cheras both, but was ultimately killed in the battle. After the third century A.D. the Chola dominion rapidly declined as a result of repeated attacks by the Pāṇḍyas and the Cheras and subsequently by the Pallavas. The fortunes of the Cholas suffered a serious setback, when a good part of the port town of Puhar was engulfed by the sea in terrific tidal waves during the reign of the later king Killivalavam. From the fourh to the ninth century A.D. the Cholas played only a marginal part in South Indian history.[5]

2. *The Cheras* : The Chera or Kerala kingdom was the Western coastal strip above the northern limit of the Pāṇḍyan kingdom. It had a number of good ports, Ṭoṇḍi, and Musiri or Muziris being the best known. The capital of the Chera kingdom was Vanji, though its location has been the subject of an inconclusive debate: some identifying it with some place on the Periyar river or at its mouth, others locating it inland in Karur or Karuvur, the centre of the

---

4.　　K.A. Nilakanta Shastri, *Op. Cit.*, pp. 104-105.
5.　　*Ibid.*, pp. 106-108.

Western-most taluq of Tiruchirapalli.[6] In the early centuries of the Christian era, the Chera country was as important as the country of the Cholas. It owed its importance to trade with the Romans. The Romans set up two regiments at Muziris identical with Cranganore in the Chera country to protect their interests. It is said that they built there a temple of Augustus. One of the earliest and better known Chera rulers was Udiyanjeral (C. 130 A.D.)

The most important event in the political history of the Cheras was their fight against the Cholas about 150 A.D. In the battle, although, the Cheras killed the father of the Chola king Kārikāla, the Chera king also lost his life. Later the kingdoms temporarily came to terms and concluded a matrimonial alliance. But after the second century A.D. the Chera power declined, and we have nothing of its history again till the eighth century A.D.

3. *The Pāndyas* : The Pāndya kingdom occupied the extreme south and included the modern districts of Tirunelveli, Madurai and Ramnad, besides south Travancore. It had its capital at Madurai, which was the Tamil word for Mathura. The Pāndyas are most famous for patronising the poets and scholars of the Tamil Sangams. The early history of the Pāndyas is not very much clear, however, they are mentioned in the Aśokan edicts and in the *Rāmāyana* and the *Mahābhārata*. The earliest known Pāndyan king was Palyagasalai Mudukudumi who is mentioned in the Sangam literature as a great conqueror, a patron of poets and performer of many sacrifices.[7] Under the Pāndyas Madurai and the port Korkai were great centres of trade and commerce. The Pāndyan dominion was very wealthy and prosperous on account of the brisk Indo- Roman trade. The Pāndyan rulers also sent their embassies to the Roman emperors.

The three states were finally uprooted by the Turks in the thirteenth century A.D.

After the death of Aśoka, the Mauryan empire began to disintegrate. The political unification of India made by Aśoka could not be retained by his successors. Consequently the Mauryan empire

---

6.   R.C. Majumdar (Ed.), *The Age of Imperial Unity*, Bombay, 1953, pp. 232-233.
7.   D.N. Jha & K.M. Shrimali (Ed.), *Prāchina Bhārata Kā Itihāsa* (Hindi), Delhi, 1986, pp. 259- 261.

was divided into different parts. The southern part of the empire was captured by the Sātavāhanas whose kingdom was extended upto western India, and their centre was at Pratisthān (modern Paithān in Maharashtra). Since the Sātavāhanas were also known as Andhra, it is suggested that they originated in the Andhra region whence they moved westward. But according to another view they originally belonged to the western Deccan and gradually extended their territorial jurisdiction to the eastern coast and in course of time they were called Andhra. But there is no consensus among the scholars about when exactly the Sātavāhanas came into power.[8] R.G. Bhandarkar, D.R. Bhandarkar, H.C. Raychaudhury and D.C. Sircar are of opinion that the rise of Sātavāhana power should be placed during the second quarter of the first century B.C.[9] V.A.Smith and A.S.Altekar, on the other hand opined that the rise of the Sātavāhana dynasty should be placed in the last quarter of the third century B.C. [10] K.A.Nilakanta Shastri is also in agreement with this view.[11] Traditionally the first ruler of this dynasty was Simuka(C.230 B.C.) who reigned for twenty-three years and it is believed that he had destroyed the Śuṅga power.[12] So, it may be surmised that the Sātavāhana came into power in about third century B.C. The earliest Sātavāhana king who received recognition was Sātkarni (27-17 B.C.), Sātkarṇi is said to have expanded his empire in all directions and received the epithet 'Dakṣiṇāpathapati' (Lord of the southern regions). A century after Sātkatṇi, the Sātavāhana power was eclipsed due to the wave of Scythian invasions and remained confined to the east of Deccan.[13] Gautamiputra

---

8.   *Ibid.*, p. 233.

9.   For detailed discussion, see A.S. Altekar, "when did the the Sātavāhana Dynasty Begin", *Proceedings of the Fifteenth Indian History Congress*, Gwalior, 1952, p. 40.

10. *Ibid.*, pp. 40,44.

11. K.A.Nilkanta Shastri, *Op. Cit.*, p. 78.

12. R.C.Majumdar(Ed.), *Op.Cit.*, p. 195; A.L.Basham, *The Wonder that was India*, Delhi, 1981, p. 62.

13. R.C.Majumdar(Ed.), *Op. Cit.*, p. 101-113; R.Thapar, *A History of India*, Vol.I, Penguin Books , 1972, pp. 99-104; K.L.Hazra, *Royal Patronage of Buddhism in Ancient India*, New Delhi, 1984, p. 112; R.C.Majumdar, *Ancient India*, Delhi, 1974, p.134.

Satkarni(106-136 A.D.) restored the Sātavāhana power and extended it upto Rājputānā and Gujarat.[14] Vasiṣṭhaputra Satkarṇi, (C. 130-159 A.D.) also known as Vasiṣṭhiputra Pulumavi is said to have had consolidated the Sātavāhana empire.[15] Among the later Sātavāhana ruler, Yajñasri Satkarṇi (C.174-203 A.D.) was a powerful ruler. The Sātavāhana empire by the first quarter of third century A.D.[16] began to decline, consequently a number of independent kingdoms came into existence. The Abhiras snatched away Maharastra from the Sātavāhana territory. In the north Kanara District of Mysore, Kuntala and Chutu became powerful and after that Kadambas became powerful in the same region. The Ikshavākus established themselves in the Andhra country. In the southeast region the Pallawas established their separate kingdom whereas in the middle of sixth century A.D. They became a great power. The Vākāṭakas also established themselves in the Vidarbha region.[17]

The Ikshavākus, who under Vasiṣṭhiputra Sāntamula, occupied the Andhra country, had their capital at Vijayapuri, modern Nāgārjunakoṇḍa, Sāntamula is credited to have performed the Vedic sacrifices viz., *Aśvamedha* and *Vājpeya*.[18] King Sāntamula was succeeded by his son Māṭhariputra Virapurushadatta who ruled about twenty years which may be roughly assigned to the third quarter of the third century A.D. Records of his reign have been found at the Buddhist sites of Amarāvati, Jaggayyapeṭā and Nāgārjunakoṇḍa. The inscriptions record the private donations to some Buddhist establishments. Nāgārjunakoṇḍa inscriptions are mostly records of the charitable gifts of some female members of Virapurushadatta's family to the cause of Buddhism who were devout Buddhists by faith.[19]

---

14.   K.L.Hazra, *Op. Cit.*, pp. 173-176.

15.   B.R.Prasad, *Art of South India-Andhra Pradesh*, Delhi, 1980, p. 6.

16.   D.N.Jha & K.M.Shrimali (Ed.), *Op. Cit.*, p. 235.

17.   *Ibid.*

18.   K.A.Nilakanta Shastri, *Op. Cit.*, p. 83.

19.   R.C.Majumdar, *Op. Cit.*, p. 225; D.C.Sircar, *Select Inscriptions*, vol.1,Calcutta, 1942, pp. 219-228.

Virapurushadatta was succeeded by his son Ehuvula Sāntamula II who ruled about eleven years. By the end of third century A.D., the Ikshavāku rule came to an end and they were supplanted by the Pallavas.

The origin of the Pallavas is shrouded in mystery. However, it is said that they were a local tribe. The father of Sivaskandavarman is regarded as the first Pallava king whose accession is dated towards the beginning of the second quarter of the third century A.D.[20] Simhavishnu Avanisimha, who ascended the throne in about 575 A.D. is considered as the real founder of the Pallava dynasty. The Pallavas made Kānchi their seat of authority which was at one time a prominent centre of Buddhism in South India and produced a number of erudite Buddhist scholars like, Āryadeva, Dinnāga and Dharmapāla etc.[21] Simhavishnu was succeeded by his son Mahendra-Varman I.[22] He was succeeded by Narasimha-Varman Mahāmalla(C.630-668 A.D.) who is considered as one of the greatest rulers of the Pallava dynasty and his political achievement made him supreme in South India.[23] It was during his period when Huien-Tsang visited Kānchi about 640 A.D. He mentions that there were more than one hundred Buddhist monasteries housing over 10,000 monks of the Sthaviravādin school.[24] He also mentions that though Buddhism was in decaying stage in South India but its position in Tondāmandalam was conspicuous. Narasimha-Varman was succeeded by his son Mahendra-Varman-II who ruled for a brief period from circa 668 to 670 A.D. He was succeeded by Parameśvara-varman I who came to the throne and who ruled from circa 670 to 695. A.D.

By the end of the eighth century A.D. the power of the Pallavas began to weaken on account of the repeated invasions by the Rāstrakutas and the internal dissention etc. One of the last great Pallava rulers was Nandivarman II (circa 730 to 800 A.D.) who had to fight against the Pāndyas and Rāstrakutas. At the close of the

---

20. K.R.Subramanian, *Buddhist Remains is Andhra...* Madras, 1932, pp. 76-77.
21. R.C.Majumdar (Ed.), *The Classical Age*, Bombay. 1954, p. 258.
22. S. Sengupta, *Buddhism in the Classical Age*, Delhi, 1984, p. 107.
23. R.C.Majumdar (Ed.), *Op. Cit.*, p. 260.
24. T.Watters, *On Yuan Chwang's...* Vol.II, London, 1905, p. 226.

ninth century A.D.(circa 893 A.D.) the Chola ruler Āditya who was a feudatory of the Pallavas defeated the last Pallava ruler Aparājita and killed him. Thus the Pallavas were supplanted by the Chola of Thanjavur in the far south.

BRIHATPHALĀYANAS : After the fall of Ikṣāvākus, the eastern Andhra region, parts of Kriṣṇa and Guntur districts came in the hands of Brihatphalāyanas in the first quarter of the fourth century A.D. Only one king of this dynasty Jayavarman, is known from a Koṇḍāmudi copper plate grant. He is described as a devout worshipper of Maneśvara.[25]

ĀNANDA : On the basis of epigraphical evidences, it appears that Ānanda dynasty rose into power from the ashes of the Ikshavākus in the later part of the fourth century A.D. Only three kings of this dynasty are known viz., Koṇḍara, Aṭṭivarman and Dāmodaravarman. There is difference of opinion among scholars regarding the dynastic name and chronology of the Ānanda kings. According to one opinion it is called Kandara family and according to other, the Ānanda-gotra family. They were followers of Brāhmanism and Buddhism.[26]

SĀLAṄKĀYANAS : The Sālaṅkāyanas ruled over Kriṣṇa, Guntur and Tenali regions after the Brihatphalāyanas.[27] They were the followers of Brāhmanism. Nandivarman-I, Chandravarman, Nandivarman II etc. were the known kings of this dynasty. It is said that the Sālaṅkāyanas were subdued by Viṣṇukuṇḍins in the beginning of 6th century A.D.[28]

VIṢṆUKUṆḌINS : This dynasty was founded by Mādhavavarman I in about 440 A.D. and seven rulers of this dynasty for three hundred years. It appears that the kings of this dynasty were tolerant towards all religions and gave their patronage to Buddhism, whereas, by faith they were Śaivites or Vaiṣṇavites. The Visṇukuṇḍin dynasty was uproored by the Chālukyas of Badāmi, when Pulakeśin II of this dynasty put an end to their supremacy in about 634 A.D.[29]

---

25.   R.C.Majumdar(Ed.), *Op. Cit.*, pp. 226- 227.
26.   *Ibid*.
27.   *Ibid*., pp. 204-206.
28.   *Ibid*, p. 206.
29.   *Ibid*., pp. 208-211.

7.2 : Buddhism in Andhradeśa:

Buddhism held a place of considerable importance in Andhradeśa and adjoining portion of South India, But due to the lack of adequate evidence it is very difficult to determine when exactly Buddhism was introduced for the first time in South India. Some later Buddhist works mention that the Buddha himself visited South India. Huien Tsang mentions a large number of places in South India associated with the memory of Buddha's preaching.[30] The Pāli text, the *Dhammapada Aṭṭhakathā* mentions one of the previous births of the Buddha was in Amarāvati. [31] One of the Tibetan traditions mentions that Śākyamuni had preached the Kālacakra system in Dhānyakaṭaka.[32] In the Vajrayāna works it is mentioned that the Buddha had turned the third wheel of law at Dhānyakaṭaka sixtreen years after His enlightenment.[33] But all these traditions are not corroborated with the archaeological evidences, however Buddhism was known in Andhradesa'a by the fourth century B.C. in the pre-Aśokan period as suggested by Amarāvati finds.[34]

As it is well known that in the second Buddhist council, held at Vaishāli, the Vajjians or Vajjiputtakas who refused to accept the decision of the council and convened another council under the leadership of Mahādeva, is said to have actually come from Andhra.[35] Moreover, the *Cullavagga* mentions that there was one monk named Puraṇa who had refused to accept the law at the first Buddhist council because he wanted to stick to the Dhamma and the Vinaya as he himself had heard from the Master.[36] This indicates that there was a revolt in the Buddhist Saṅgha even during the time of the first Buddhist council. This revolt had also come from the Vajjiputtaka

---

30. T.Watters, *Op. Cit..* p. 209.

31. *Dhammapada Aṭṭhakathaā*, I, 83.

32. Charles Eliot, *Hinduism and Buddhism*, Vol.III, London, 1968, p. 386; P.S.Sastri, "The rise and Growth of Buddhism in Andhra," *IHQ*, Vol. XXXI, 1955, Calcutta, p. 70.

33. P.S.Sastri, *Op. Cit.*, p. 70.

34. B.R.Prasad, *Op. Cit.*, p. 3.

35. P.S.Sastri, *Op. Cit.*, p. 71, according to another opinion Mahadeva hailed from Mathura, cf. N.Dutta, *Early Monastic Buddhism*, Vol.II, Calcutta, 1941,p. 41.

36. *Cullavagga*, XI.i.II; *DPPN*, Vol. II, p. 237.

group which included Ikshākus to which the monk Puraṇa belonged and before the rise of the Sātavāhanas they were at Bhaṭṭiprolu and later moved towards Nāgārjunakoṇḍa.[37]

The off-shoot of the Mahāsaṅghika, the Siddhatthikas the Bahusutiyas, the Rājagirikas, the Apara Śaila and the Purva Śaila and the Āryasaṅgha made their strongholds in Andhra.[38] The *Kathāvathu*[39] also mentions the name of Andhaka a sub-sect of the Mahāsaṅghika school. The only sub-sect of the Theravāda school which developed in South India was the Mahisāskas.[40]

The *Mahāvaṁsa* mentions that after the third Buddhist council, Buddhist missions were sent to different parts of India and abroad. Mahādeva went to Mahisamaṇḍala and Rakkhita to Vanavāsi, both these countries have been located in South India.[41] Buddhism received a boost in South India during the Aśokan period. Aśoka in his 3rd Rock-Edict mentions that his *Dhamma Vijaya* prevailed in the border kingdoms of the Choḍa (Cholas). Pāḍa(Pāṇḍayas) and as far as Tāmbapanni(Ceylon). But it was his son Mahinda who was more responsible for the spreading of Buddhism in South India. It is said that in this connection he got help from Mahā-Ariṭṭa, a nephew of the Ceylonese king Tissa. Mahinda is said to have erected seven *Vihāras* at Kāveripattinam, while he was on his way to Ceylon.[42] During the time of Aśoka the Mahāsaṅghika sect gained a firm footing there.[43] Some of the *Stūpas* at Andhra may be counted among the thousands erected by Aśoka all over India.[44]

The Mahāyāna Buddhism also established their position in South India from the very time of their appearence. Nalinaksha Dutta is of opinion that Mahāyāna Buddhism originated in Andhra country.[45] It

---

37. P.S.Sastri, *Op. Cit.*,

38. *EP.*, Vol, XX, Nāgārjunakoṇḍa Inscriptions.

39. *Point of Controversy*(PTS), p. 104.

40. *Dipavaṁsa* (Ed. by Oldenberg), Chapter IV.

41.   B.N. Chaudhury, *Buddhist Centres in Ancient India*, Calcutta, 1982, p. 229.

42. A.Aiyappan and P.R.Srinivasan (Ed.), *Story of Buddhism with special reference to South India*, Govt. of Madras, 1956, p. 53.

43. B.R.Prasad, *Op. Cit.*, p. 4.

44. K.R.Subramanian, *Op.Cit.*, p. 4.

45. R.C.Majumdar(Ed.), *Age of Imperial Unity*, Bombay, 1953, p. 388.

is interesting to note here that a passage occuring in the *Aṣṭasāhasrikā Prajñāpāṛmitā* states that "Mahāyāna Buddhism will originate in the Dakṣiṇāpath(Southern India), and pass into the North.[46] From the epigraphic evidence we learn that the Mahāyāna Buddhism gained their wide spread popularity in South India, but the Theravāda Buddhism was more popular in this region in the early centuries of the Christian era.

The Sātavāhanas also patronised Buddhism during their priod; we saw immense development of Buddhism and its art during their period. The Sātavāhana ruler Gautamiputra Sātkarni and his mother were also favourable towards Buddhism as evidenced by numerous inscriptions and Bhadrayānika seems to be an important sect in his kingdom.[47] Inscriptional evidence suggest that Buddhisim attained great prominence during the reign of king Vāsiṣṭhiputra Pulumāvi. During this period Buddhism was not only patronised by the members of the royal family but also by the common people. [48] Although Yajñasri Sātkarni was the follower of Brāhmanical religion but he revered and patronised Buddhism. During his period several Buddhist edifices were also erected.

After the Sātavāhanas, the Ikshavākus came into the power in the Andhra country.[49] During this period Buddhism flourished more vigorously and gigantic monasteries and *Stūpas* of exquisite craftsmanship of architectural and sculptural grandeur were constructed throughout the country with the liberal munificence of the royal members and the feudatory chieftains. Though the first ruler of this dynasty, Vāsiṣṭhiputra Shri Sāntamula[50] was follower of Brāhmanism and during this period Brāhmanism received a boost but still Buddhism was very much prevalent; Vāsiṣṭhiputra Shri Sāntamula was succeeded by his son Māṭhariputra Shri Virapurusadatta who was a great patron of Buddhism as suggested by numerous inscriptions found at Amāravati, Nāgārjunakoṇḍa, Jaggayyapeṭā and other centres.[51] The

---

46. *Ibid.*, p. 387.
47. K.L.Hazra, *Op. Cit.*, pp. 172-173.
48. *Ibid.*, pp. 173- 176.
49. R.C.Majumdar, *Op. Cit.*, p. 224.
50. *Ibid.*
51. K.L.Hazra, *Op. Cit.*, pp. 180-193.

royal ladies and female members of other Kinsmen were among the chief donors of the Buddhist Saṅgha.[52] The foremost donor among them was the Princess Mahātalvari Mahāsenapatni Mahādānapatni Sānti Shri, a . paternal aunt and mother-in-law of king Shri Virapurusadatta. Her name is mentioned for her benevolent acts in the inscriptions engraved in several *Āyaka* pillars at the *Mahācaitya* on the Śriparvata.[53] In the sixth year of the reign of Shri Virapurusadatta she re-erected the *Mahācaitya* and the *Mahāvihāra* on the Śriparvata and set up *Āyaka* pillars in each of the four cardinal directions, as mentioned in one of the *Āyaka* pillar inscriptions.[54] Sānti Sri had also erected apsidal temples, monasteries, stone halls and set of rooms for the use of the Buddhist teachers of Śriparvata.[55] It is said that she was a disciple and adherent of the teachers of the Aparamahāvinaseliya. Among other prominent ladies who contributed to the construction of the Buddhist edifices etc. are Mahātalvari Sānti Shri, Mahāsenāpatni Cūla Śāntisiriṇikā, Skandacalikiraṇaka etc.[56] Sri Virapurusadatta was succeeded by his son Vāsaṣṭhiputra Shri Bahubala Sāntamula. Only a few material is available regarding the condition of Buddhism of his time. Two inscriptions record the name of royal ladies who made donations to the Buddhist Saṅgha.[57]

The Ikshavākus were succeeded by the Pallavas of Kānchi and the Chālukya of Badāmi respectively who were the followers of Brāhmanism and this lack of patronage and also some short of apathy led Buddhism to the path of decline.[58]

The *Silappadikaran* and the *Manimekālai*, two Tamil classical works contain references to Buddhism. But *Silappadikaran* does not deal with Buddhism or any other religion in particular. The only clear reference to Buddhism can be found in Kovalan's narration of his dream to the Brāhmaṇ Madalan in which Kovalan says that he saw Mādhavi surrendering her daughter Manimekālai to a life of asceticism.[59] A

---

52. *Ibid.*, pp. 180-181.
53. B.N.Chaudhury, *Op. Cit.*, p. 229.
54. *EI*, Vol. XX, p. 19.
55. *Ibid.*, Vol. XXI, pp. 64-66.
56. B.N.Chaudhury, *Op. Cit.*, pp. 230-231.
57. *Ibid,.* p. 232.
58. K.L.Hazra, *Op. Cit.*, p. 197.
59. A.Aiyappan & P.R.Srinivasan (Ed.), *Op. Cit.*, p. 53.

reference to an Āriya or Palli in Puhār is to a Buddhist shrine.[60] The *Manimekālai* on the other hand is a great Tamil work on Buddhism, written by Settlai Sattanar in the second century A.D. The story of the conversation and activities of Manimekālai as a nun is narrated in this epic poem. It is said that Arāvaṇa Adigal, a Buddhist teacher had converted Manimekālai and helped in her mission of helping humanity.[61] The *Manimekālai* also mentions about the prevalence of Buddhism in Kerala. The keralolpathi, the earliest traditional account of Kerala, mentions about twenty-five Perumals and gives details about the administration of each one of them. According to it two of the Perumals named Palli Bana Perumal and Cheraman Perumal embraced Buddhism.[62] But what form of Buddhism was prevalent in Kerala is not known.

The Sālankayānas who were the contemporary of the Guptas, occupied the position next after the Ikṣvākus for their patronage to Buddhism. Though they were adherents of Bhāgavatism but adopted a tolerant religious policy. It is said that they had contributed to the propagation of Buddhism in Burma.[63]

The period of Kalabhra rule in South India seems to have been favourable to Buddhism. Under the patronage of Accyutavikkanta, Buddhadatta wrote many books. In the *Abhidhammāvatāra* he gives a glowing account of Kāveripattinam and many monasteries.[64] It is also said that he held the charge of the monasteries at Kāveripattinam, Uragapuram, Bhutamangalam and Kānchipuram etc.[65],

The Buddhist philosopher, Bodhidharma, the founder of Dhyāna-Buddhism lived early in the 6th century A.D. and was the prince of Kānchipuram. It was he who was responsible for introducing this faith in China from where it spread to Japan. This faith is called 'Chān' by the Chinese and 'Zen' by the Japanese. It is said that Mahāyānic philosophers and commentators like Āryadeva,

---

60. *Ibid.*
61. B.N.Chaudhury, *Op.Cit.*, pp.231-232.
62. *Ibid.*
63. *Ibid.*, p. 234.
64. M.Tiwary(Ed.), *Abhidhammavatar*, Delhi, 1987, pp. 19ff.
65. A.Aiyappan & P.R.Srinivasan(Ed.), *Op.Cit.*, p. 54.

Buddhapālita, Diṅnāga and Bhāvaviveka were living in South India.[66] Buddhaghoṣa was patronised by Saṅghapāla, a king of Kāñchipuram, Dharmapāla is said to be head of a Buddhist monastry of Bhataraditta *Vihāra* at Kāñchipuram.[67]

In the 7th and 8th century A.D., Buddhism had a strong opposition from Śaivism, The form of Buddhism prevalent during this period was Tantrayāna.

The Chinese travellers Huien Tsang and I-tsing give us a graphic account of the state of Budhism of this period. Huien-Tsang says that he found ruins of the monasteries built by Aśoka and his son Mahinda in South India.[68]

I-tsing says that Nāgapattinam was the chief centre of Budhism.[69]

Buddhism lingered in South India at least upto the period of 14th century A.D. After that the Buddhist who remained in South India were gradually converted into Hinduism.[70] But many Buddhists kept Buddhism alive until the 16th century A.D. who used to stay at Nāgapaṭṭinam.[71]

7.3 : Andhra school of Buddhist art :

The Andhra school of Buddhist art was one of the most prolific schools of Indian art. The dating of this school is not beyond controversy. The reason for such a split in opinion amongst scholars is probably due to the placing of too much emphasis either on the palaeography of short legends inscribed on pillars and other sculptured panels or on the stylistic traits of carved sculptures. However, generally scholars are unanimous to fix the date of this school of art either in the last quarter of 2nd century B.C. or at the beginning of the last century B.C.

---

66. Tāranāth, *History of Buddhism in India*(Tr.) by Lama Chimpa and Alka Chattopadhyaya, D.P. Chattopadhyaya(Ed.) Simla, 1970, p. 277.

67. A.Aiyappan & P.R.Srinivasan (Ed.), *Op. Cit.*, p. 55.

68. T.Watters, *Op. Cit.*, pp. 214-215.

69. A.Aiyappan & P.R.Srinivasan, *Op. Cit.*, pp. 55-56.

70. *Ibid..* pp. 57-58.

71. *Ibid.*, p. 58.

The subject matter of the early phase of Buddhist art of South India are symbolic in character and narrative in nature as in the case of Bharhut and Sānchi etc. The anthropomorphic images of the Buddha appeared in South India in the second-third century A.D. , but it did not immediately reject the firmly established aniconic tradition. It was as if the human figure of the Buddha served new purpose and so could not entirely replace earlier symbolic representation, which disappeared only slowly.[72] It seems that a large number of Buddhist sects, as mentioned in the Amarāvati and Nāgārjunakoṇḍa inscriptions remained faithful to the earlier traditions, were somewhat reluctant to accept the human figure of the Buddha.[73] It is notable here that on the same monument or sometimes in the same relief symbols and anthropomorphic images of the Buddha are represented side by side. "The explanation for the obvious co-existence of the two different modes of iconographies lies equally in the slow evolution from aniconic to iconic forms and in a kind of routine attachment to the earlier artistic formula."[74]

The images of the Buddha which are both in relief and round, are unearthed in Jaggayyapeṭṭā, Bhaṭṭiprolu, etc. besides Amarāvati and Nāgārjunakoṇḍa.

In Andhradeśa, particularly in the lower Kriṣṇa and Godavari valley, the anthropomorphic images of the Buddha appeared between second and third century A.D. The early phase of the Buddha figures in Andhra art are mainly in *Abhaya-mudrā*, but simultaneously the other *mudrās* e.g. preaching pose etc. also occurs.[75] The early images of the Andhra school of art are influenced by the adeals of the Mathura school. On both the seated as well as standing images, the impact of Mathura is unmistakable. The figures are, no doubt, carved in set pattern, along with the physical features of a *mahāpuruṣa* and so also the fixed positon of hands and legs to denote set postures which had already been codified in earlier centres like Mathura and Gandhāra. Even the mode of wearing a coarse and heavy drapery of

---

72. D.L.Snellgrove (Ed.), *The Image of the Buddha*, Delhi, 1978, p. 78.

73. *Ibid*.

74. *Ibid*., p. 76.

75. A.Aiyappan & P. R.Srinivasan, *Op.Cit*., p. 64.

the Kuṣāṇa Buddhas has been adopted in Amarāvati art also. "It appears that this indication of Kuṣāṇa-influence on the Buddhist images was slowly descending down to South India from the 2nd century A.D. and reached the Kriṣṇa region about the construction of the rails."[76] But it does not mean that Andhra sculptures are simply a copy of the Mathura school of art. Actually the images of the Buddha from Andhra are more alive and nearer to the spirit of the Master. O.C.Gangoly aptly said, "even some of the mutilated faces convey to us a gravity, a remoteness, and spiritual quality of introspection which is far above the snug complacency which characterize the happy faces of the above the Mathura school and the banal objectively of the Gandhāra- masons,"[77] The art of stone-carving has now attained full maturity and the sculptor skilfully manages to distribute the balance between the images and the crowding scenes, do not loose overall harmony and pleasing effect. However, a few of the carved dome-slabs which is assignable to 3rd century A.D. reflect a downward trend in art. Though, the tradition of mature Amarāvati style contains yet a certain amount of rigidity and formulization is the portrayal of human figures with unduly eloquent, and somewhat stylized forms are clearly discernable.

The Chandavaram(in Prakasam district) was another important centre of the Buddhist art sometimes during the 1st century A.D. The carved sculptural panels brought out from the excavations would reveal that the Chandavaram *Stūpa* continued to flourish till the 3rd century A.D. The striking features of this *Stūpa* is the use of buff-coloured sandstone and not the Palnad type of creamish lime stone used in various other Buddhist centres of South India. The technique of Amarāvati of delineating the themes in different planes has been maintained here also. The treatment of male and female figures, the drapery and ornaments in the carved stone-panels of earlier phase at Chandravaram are very much similar or nearer, to the second phase of Amarāvati sculptures.

---

76. T.N.Ramchandran, *Buddhist sculptures from a stūpa near Goli Village*,Madras, 1929 (*ASI*).

77. O.C.Gangoly, *Andhra Sculptures*, Hyderabad, 1973, p. 52.

Goli, another Buddhist site in South India, yielded a number of Buddhist images. The images are very much similar to the sculptures of Amarāvati in theme and execution. Apart from the *Jātaka* stories, the events of the life of the Buddha are also depicted.

Stylistically, the human figures, dress, ornamentation etc. occuring on the friezes from Goli do very much resemble to some of the sculptures discovered at Mathura. However, the delineation of themes by separating different schemes with the help of three knobs or rivet heads is common to Goli and to the last phase of Amarāvati art.

Nāgārjunakoṇḍa also yielded a good number of magnificent images. It came to limelight under the Ikshvākus who had held a portion of the kingdom of the Sātavāhanas some time in the second quarter of third century A.D.[78] The themes of Buddhist art which has been adopted at Nāgārjunakoṇḍa are similar to Amarāvati e.g. depiction of *Jātakas* symbols, events of the life of the Buddha etc. Even the continuous narrative method is also been manifested here.[79] This is why it is said that Nāgārjunakoṇḍa is the missing link of the Amarāvati art.

Stylistically the art of Nāgārjunakoṇḍa can be divided into two groups. The earlier of them dating back to the initial years of Māṭhariputra Virapurushadatta consists of drum-slabs, and a few memorial pillars etc.[80] The second group of Nāgārjunakoṇḍa art may be assigned to the eighth regnal year of Ehuvala Chāṁtamula.[81] The art of early phase of Nāgārjunakoṇḍa are somewhat rigid and lack of the maturity of the Amarāvati ideal whereas the specimens of later phase exhibit an improved artistic vision and elegance.[82] The artists have tried to show the different moods and human expression. The exultant Mithuna figures are rapt in dalliance.[83] A panel discovered at Nāgārjunakoṇḍa, depicts the blushful mood of a lady to accept

---

78. H.Sarkar & B.N.Mishra, *Nāgārjunakoṇḍa*, ASI, Delhi, 1980 (reprint), p. 13.
79. *Ibid*, pp. 45-46.
80. *Ibid*., pp. 45
81. *Ibid*.
82. *Ibid*.
83. *Ibid*.. pl.XII, A.

wine being offered by the sporting spouse, is regarded as one of the superb artistic quality of Nāgārjunakoṇḍa sculptures.

The anthropomorphic images of the Buddha in Nāgājunakoṇḍa are mostly in *Abhaya* and *Vyākhyāna mudrās*, however, the *Dhyāna mudrā* is not altogether absent. The drapery is transparent which covers both the shoulders, however in a few images the right shoulder is exposed. The hanging drapery is carefully arranged in well-regulated folds as in the case of Mathura images. The eyes are usually half closed and the face is round with parted lips. The hair is arranged in curly manner.

The sculptors of Nāgārjunakoṇḍa were expert in producing the ideal feminine beauty. The female forms both at Amarāvati and Nāgārjunakoṇḍa are elongate, slim,graceful and sublime. But the art is less sensuous in comparision to Mathura school where the protruding breast and high hips of the ladies look predominent and aggressive.

The Andhra school of art seems to have begun to eclipse in the 5th century A.D. due to the growing power of the Pallavas who were the supporters of Brāhmanism.[84] But the making of images of the Buddha did not finally eclipse and they were made continuously even after the eighth century A.D.[85]

With the rise of the Mahāyāna Buddhism in South India, several images of the Mahāyānic deities also appeared in Andhradeśa. A small torso of Avalokiteśvara discovered in the Kriṣṇa valley belongs to the early period of Pallava art,[86] which is very much interesting in its style. A beautiful gilt bronze statuette of Maitreya discovered at Melayur, Tanjor district, is attributed to the eighth or ninth century A.D.[87] The statuette is richly dressed with its garment draped in a very special manner.

---

84. D.L.Snellgrove(Ed.), *The Image of the Buddha*, Delhi... p. 124.
85. A.Aiyappan & P.R.Srinivasan, *Op. Cit.* pp. 92-93.
86. D.L.Snellgrove, *Op. Cit.*, pl. 88.
87. *Ibid.*, pl.90.

The arrival of the Vajrayāna Buddhism in South India, particularly at Salihundam, north of Andhradeśa, led to the appearance of the images of Tārā, Mañjusri and Mārici etc.[88]

7.4 : Iconography of the Buddha images :

The anthropomorphic images of the Buddha appeared in South India between 2nd and 3rd century A.D. and they are mainly from Amarāvati and Nāgārjunakoṇḍa. Most of them are life-size, however, there are also a few small figures.[89] The credit for the introduction of Buddha image in Andhra school is generally given to the Chaityaka sect, an off-shoot of Mahāsaṅghikas.[90] The early images of the Buddha from Amarāvati show certain heaviness like Mathura school. The influence of Hellenism is also traceable in the art of Veṅgi or Andhra, may be due to trade relation between them. In course of time the Andhra school acquired a truely classic beauty, leaving aside the features of Mathura and Gandhāra.[91] The face of the Buddha in Andhra school resembles the tradition of Roman portraiture which has led a number of scholars to raise the possibility of Roman influence on Andhra school and which can not be completely denied.[92] But all the images of the Buddha are not influenced by the Roman art. Any way, stylistically the images of the Buddha of Andhra school can be divided into two sections: A and B.

SECTION-A : The early images of the Buddha belonging to the period before fourth century A.D. have some specific features in comparison to other schools of art:[93]

1. The face of the Buddha is narrow and oval in contrast to the round facial contour of Mathura style.

2. The body is tall and slender with broad shoulders and slender legs.

---

88. *Ibid.*, p. 128.

89. A.K.Coomaraswamy, *History of Indian and Indonesian Art*, London, 1927, fig. 137.

90. B.R.Prasad, *Art of South India-Andhra Pradesh*, Delhi, 1980, pp. 37-38.

91. Stella Kramrisch, *Indian Sculptures*, Delhi, 1981, p. 48.

92. D.L.Snellgrove, *Op. Cit.*, p. 76.

93. S.K.Saraswati, *A Survey of Indian Sculptures*, Calcutta, 1957, p. 18; D.L.Snellgrove, *Op.Cit.*, pp. 79-81.

3. The hair of the images are very short and curly twisted in regular spirals and the *Uṣniṣa* is reduced to a low conical protuberance.

4. The *Ūrṇā* is very clearly marked between the eyebrows.

5. The drapery has five pleats and generally covers the left shoulder only but in a few images both the shoulders are covered like Gandhāran images, which mostly occur in the bas- reliefs.

6. The over-garment (*Uttarāsaṅga*) is very long and is held in the crook of the left arm which bends towards the shoulder and falls in a broad pouch to ankle level.

7. The bottom edge of the lower garment (*antaravāsaka*) is visible and thrown over the back and hangs free.

8. Though, the style of the garment appears to have derived inspiration from Mathura school, but there are quite some clear differences. The drapery is not transparent and it does not cling so closely to the body which clearly shows the privy parts.

The visible belt in the manner of typical style is totally absent. In fact the garment of the images hangs loose which does not show the inner portion of the body.

9. The figures are mostly in *Abhaya-mudrā* but the gesture of left hand is slightly changed from the Mathura style. The variations in style are most discernible in the bas-reliefs.

10. In contrast to Mathura or any other school of art the figures of the Buddha do not show in *Vajrāsana* with their legs tightly crossed and the soles of the feet upturned. The legs are simply folded ones above and in initial period the knees are always very far apart and the ankles touch each other.

11. Some images of the Buddha are shown seated in *Bhadrāsana* or *Pralambapādadsana*.

Some of the specimens which bear above mentioned features are as follows :

1. A Buddha head discovered at Vijiaderpuram, and housed in 'Musee Guimet', Paris, is considered as one of the best specimens of the Andhra school. The eyes of the image are open and the *Ūrṇā* between the eyebrows is prominent. The nose and the left cheek of

the image are partially damaged. In contrast to Mathura and Gandhāra schools, the hair is shown here neatly coiled around the head in curb that covers cranium and the small *Uṣniṣa*. This style of the hair would seem to be the point of departure for a stylization that was adopted by all schools of Buddhist art. The face shows affinity with the traditional Roman portraiture as it is well-known that there were trade contacts between South Indian states and Roman world so the Roman impact on Andhra sculptures can not be refuted.[94]

2. A standing image of the Buddha from Amarāvati, housed in the Government Museum, Madras, is the next representative of this section.[95] The right hand of the image is fully damaged and the upper portion of left hand is also damaged. The *Ūrṇā* is clearly marked between the eyebrows. The face is also slightly damaged. The neatly and finely pleated garment reveals the shoulder and the right side of the chest is held in position over the bent left forearm.

3. The next representative of this section is a stone slab from the dome of a *Stūpa* from Nāgārjunakoṇḍa ,housed in Nāgārjunakoṇḍa Archaeological Museum.(Pl.22).[96] The slab depicts two scenes from the life of the Buddha. Among these two scenes, one is on the left side, in which the Buddha is shown seated in *Pralambapādaāsana* and receiving the alms from the four kings, two lions are depicted on the pedestal of the throne. *Bodhi*-tree is also depicted besides the throne. On the right side of the slab, the Buddha is shown in the same manner but receiving the alms from the two merchants who are recognised as Tapussa and Bhallika. The Buddha depicted on the left side has worn a drapery which covers the left shoulder only whereas the drapery on the right side of the Buddha covers both the shoulders. The hair of both the images has been arranged in curly manner and halo is plain.

4. A Buddha head from Vidyadharapuram, housed in Madras Museum is the next representative of this section.[97] The image is dated in the period 3rd-4th century A.D. and it is comparatively

---

94.   D.L.Snellgrove, *Op. Cit.*, pl. 47.

95.   *Ibid.*, pl.48.

96.   *Ibid.*, pl. 52.

97.   A.Aiyappan & Srinivasan, *op.Cit.*, pl. 7.

elongated. The lips are pronounced and eyes are accentuated by the incisions to demarcate the eye-lids and the curls of hair are broader. The facial expression is serene rather than meditative. The nose, ears, *Ūrṇā* and the *Uṣniṣa* are all broken.

5. The next representative of this ssction is a seated cross-legged image from Nāgārjunakoṇḍa in which the Buddha is shown seated on a throne-like pedestal under which two deers are depicted seating front to front which signifies the Sārnāth episode (Pl. 23) The folded drapery covers the left shoulder only and the right hand is turned in typical pose which may indicate the gesture of argumentation while the left hand is held akimbo on which the hem of the garment rests. The hair is arranged in curly manner with *Uṣniṣa*, the ears are elongated and the halo is plain. The *Bodhi*-tree is depicted behind the halo. The Buddha is here flanked by two deities seated cross-legged; their hands are in *Añjalimudrā*. These three deities including the Master, are flanked by four standing attendants, two on each side and among them two *chauri-* bearers flanked the Buddha closely. The pedestal of all the images is supported by four lions.

6. The next representatives of this secton are two standing Buddha images which have been shown together (Pl.24). The drapery of the images are folded and covers the left shoulders only and the lower garments reaches below the knees, both the hands of the images are missing, however, on the basis of the setting of the garments we may surmise that left hands of the images were held akimbo on which the hem of the garments rest. The head of one of the images is arranged in curly manner with *Uṣniṣa*. There is *Ūrṇā* in between the eyebrows.

SECTION-B : From the second half of the fourth century A.D. we find the abandonment of a number of Buddhist sites of which Nāgārjunakoṇḍa is more noticeable. However, other centres remained more or less active. In the sixth century A.D. Amarāvati again revived and remained an important centre of art at least upto the fourteenth century A.D. In this period the images of the Buddha are mainly in round majority of whom are in bronze. The images of the Buddha belonging to the period circa 6th century A.D. are influenced by the Gupta art, however a few features of the primitive style remain in existence as for example, the *Uṣṇiṣa* continued to be as small as those of the Amarāvati school. The characteristic features of the images of the Buddha belonging to this section are as follows:[98]

1. The face is round and the eyes are almond shaped. The lips are thick and full. The drapery is usually folded which suggests the continuity of tradition but folded drapery is now loose in appearance in contrast to early period. The folds are more shallow and they appear in conjunction with the appearance of broad ends of fabric crossing the left shoulder. In some cases the drapery is without folds which suggests some sort of a connection in this field of activity between North (particularly Sārnāth) and South India at that period.

2. The *Uṣṇiṣa* now becomes almost a top knot and the curls of the hair became more prominent.

3. The gesture of *Varada-mudrā* has become very common.

Some of the specimens which bear above mentioned features are as follows :

1. A standing bronze image of the Buddha from Buddhapada, Tamil Nadu and housed in the British Museum is one of the best examples of this section.[99] The image belongs to the period of 6th century A.D. The face of the image is oval, eyes are half closed and the ears are elongated. There is *Ūrṇā* between the eyebrows. The hair is curly and the *Uṣṇiṣa* is prominent. The left leg and the left hand are broken, the foldless drapery is clearly visible which may relate to the influence of Sārnāth school of art. The right hand is in *Varada-mudrā*. The overall face is impressive.

2. The next representative of this section is a standing bronze image of the Buddha from Amarāvati, housed in the Madras Museum.[100] The image belongs to the period 5th-6th century A.D. The drapery covers the left shoulder of the image only. The right hand is in *Varada-mudrā* whereas the left hand is holding the end of the garment. The *Uṣṇiṣa* shows protuberance and the hair is smooth and rubbed hence it is not very clear. The eyes are half-closed and

---

98. A. Aiyappan & P.R. Srinivasan. *Op. Cit.*, p. 69.

99. D.L. Snellgrove, *Op. Cit.*, pl. 87.

100. A. Aiyappan & P.R. Srinivasan, *Op. Cit.*, Pl. VI. fig. 9.

the ears are elongated. The mark of *Ūrṇā* between the eyebrows is completely absent. The type of workmanship and features are perhaps similar to the image from Buddhapada.

3. A standing image of the Buddha from Kānchipuram which belongs to the period 6th-7th century A.D. is the next representative of this section. The image is housed in the Madras Museum.[101] The head of the image is oval and the eyes are broad. The hair is arranged in wide curls and the *Uṣniṣa* and nose is very worn. The forearm of both the hands are broken and missing, however, the remaining portion indicates that the right hand was turned in the *Abhaya- mudrā* and the left hand carried an alms bowl. The figure is massive and bold. The drapery is folded and covers both the shoulders. The features of this image suggest the continuity of early tradition of South Indian art. So this proves the fact that tradition of art had a tendency to continue unbroken in the South.

4. A seated image of the Buddha from Marudurkulangarani near Trivandram is another representative of this section.[102] The Buddha is shown here in *Dhyāna-mudrā*. The face of the image is round and the drapery is folded. The image is very much similar to the above mentioned Kānchipuram Buddha image. Coomaraswamy[103] says that this image is also very much similar to the seated image of the Buddha from Anurādhapur which is dated to the period 7th-8th century A.D.

5. The next representative of this section is a seated image of the Buddha from Amarāvati.[104] The Buddha is shown here in *Bhūmisparśa mudrā* (earth-touching posture). The style of the drapery is not known because of the roughness of the stone. However it is said that there should have been the three pieces of cloth (*Trichīvara*) on the figure. The head of the image is round and the eyes are half closed. The earlobes are elongated and there are lined mark round the neck (*Trivali*). The nose is partially damaged and the lips are

101. T.A.G.Rao, "Buddha Vestings in Kanchipuram", *IA*. Vol.44, p. 127.

102. A.K.Coomaraswamy, *Op. Cit.*, p.161; T.A.G.Rao, *Travancore Archaeological Series*, Vol.II, partII, pl.III.

103. A.K.Coomaraswamy, *Op. Cit.*, p. 250, pl.XCVIII.

104. A. Aiyappan & P.R. Srinivasana, *Op.Cit.*, p. 74, fig. 10.

worn out. The *Ūrṇā* mark between the eyebrows is absent. The hair seems to be curly, now it is rubbed out, but *Uṣṇiṣa* is clearly visible. The oval halo is plain. Marks on the pedestal seems to be a lotus throne. The facial features of this images suggest the attitude of contemplation. On the basis of execution and features the date of the image is assigned to the period 8th-9th century A.D.

6. A standing image of the Buddha preserved in Bezwada Museum (Pl. 25) is the next specimen of this section. The face of the image is comparatively flat, ears are elongated, the eyes are open, there is a mark between the eyebrows which indicates previous the existence of *Ūrṇā* and the hair is arranged in curly manner with *Uṣṇiṣa*. The folded drapery covers the left shoulder only and the lower garment reaches below the knee, the extreme lower portion of the body is broken. The right hand of the image is missing while the forearm of the left hand is broken, however, the remaining portion indicates that it was held akimbo on which the hem of the garment rests.

CHAPTER - 8

ICONOGRAPHY OF THE BUDDHA IMAGES IN EASTERN SCHOOL OF ART

8.1 : *EASTERN INDIA :Historical perspective of the region:*

A. Bihar, B. Bengal, and C. Orissa.

8.2 *Buddhism in Eastern India:*

A. Bihar, B. Bengal, and C. Orissa.

8.3 : *Eastern school of Buddhist art:*

A. Bihar, B. Bengal and C. Orissa.

8.4 : *Iconography of the Buddha images:*

A. Bihar, B. Bengal and C. Orissa.

8.1 : EASTERN INDIA : Historical perspective of the region:

A. *BIHAR* : The present day Bihar is one of the states of Indian Union which is situated in eastern India in between Uttar Pradesh and West Bengal. Two of the prominent ancient *Mahājanapadas*, Aṅga, Magadha and the adjoining areas later on came to be known as Bihar because of the predominence of Buddhist monasteries (*Vihāras*) which the area was dotted with.

In Buddhist literature there is mention of Aṅgadeśa whose capital was Campā. Modern districts of Bhagalpur and Monghyr in Bihar was the part of Aṅga. It was one of the flourishing cities and a great centre of trade and commerce in Ancient India. During the period of the Buddha, there had been a rival condition between Aṅga and Magadha. It is said that for a short period Magadha had become part of the Aṅgadeśa but very soon the power of Aṅga declined and Bimbisāra taking this benefit, snatched Magadha and along with this he also incorporated Aṅga into his empire and from this period onwards Aṅga became a part of the Magadhan empire.[1]

The Magadhan empire came into prominence during the reign of Bimbisāra, a contemporary of the Buddha, who belonged to the Haryaṅka dynasty. Bimbisāra is said to have ruled for about 52 years.[2] He had pursued the policy of expansion. It is said that he possessed certain advantages denied to many of his contemporary rulers of various states. He was the ruler of a compact kingdom protected on all sides by montains and rivers and his capital at Girivraja was enclosed by five hills etc.[3] One of the important achievements of Bimbisāra was the annexation of the neighbouring kingdom Aṅga, as mentioned earlier. He had also improved his position with matrimonial alliances with the royal families of Kosala and Vaisāli. With the marriage of Kosala princess he got Kāshi village as dowry. The Vaishali marriage ultimately paved his way to the expansion of Magadhan empire to the northern region. Bimbisāra is also credited to have built the city named Rājagriha or the king's house now known as Rājgir situated near Patna, Bihar. Bimbisāra was succeeded by his son Ajātasatru.[4] Tradition says that in his old age the king was murdered by his son Ajātasatru.

Ajātasatru, also known as Kuṇika, was a conqueror. He fought many wars and throughout his reign he pursued an aggressive policy of expansion.[5] His reign saw the high watermark of the Bimbisāra dynasty. Ajātasatru had hostility with the rulers of Śrāvasti, Vaishāli, Kuśinagar and Pāvā. The territory of Vriji was ultimately annexed to the kingdom of Magadha. He had prolonged conflict with Kosala and ultimately he got the best of the war and Kosalan king was compelled to purchase peace by giving his daughter in marriage to Ajātasatru and leaving sole possession of Kashi to him.[6] It is said that Ajātasatru was the patron of Devadatta, cousin of the Buddha but later on he

---

1. R.C. Majumdar (Ed.), *The Age of Imperial Unity*, Bombay, 1953, pp. 3-4.

2. *Ibid.*, pp. 18-38.

3. R.C. Majumdar, H.C. Raychaudhury and etc.(Ed.), *An Advanced History of India*, part I, London,1949,pp.58-59.

4. *Ibid.*

5. *Ibid.*

6. *Ibid.*, p.6.

was inclined towards the Master. It is well known that the first Buddhist council, held at Rājagriha was patronised by Ajātsatru. He was succeeded by his son Udāyi or Udāyabhadra. With the merger of Kāshi, Aṅga and Vriji into the Magadhan empire, it bacame larger and consequently the seat of authority changed from Rājagriha to Pāṭaliputra. Udāyabhadra was followed by Aniruddha, Munda and Darśaka respecively. After that Śiśunāga became the king and he was succeeded by Kālāśoka.[7] It was during his reign when the second Buddhist council was held at Vaishāli.[8] Kālāśoka could not enjoy his position for a long period as he was assassinated and after that the Magadhan empire came into the hands of Nanda dynasty which was by founded Mahānandin. Mahāpadmananda was a well known and powerful king of this dynasty. He had annexed his neighbouring territories into his empire. He is said to have belonged to a *Śūdra* lineage. He was succeeded by his eight sons, respectively, among them Dhanananda was the last ruler of this dynasty who was dethroned by Chandragupta Maurya.[9] During the reign of the Nandas, Magadhan empire flourished more vigorously and it was the Nandas who made Pāṭaliputra the centre of Northern Indian power and consequently Pāṭaliputra became not only a political centre but also a centre of learning and culture.

The Mauryan empire which was founded by Chandragupta Maurya in the early fourth century B.C., ruled over the whole of Bihar. He was succeeded by his son Bindusāra who ruled about twenty four years and then Aśoka the great came into the throne who ruled about forty years. After the death of Aśoka in circa 232 B.C. the history of the Mauryan empire is shrouded in obscurity; however, it is said that after Aśoka, the Mauryan empire was divided into two parts; the eastern and the western parts. The eastern parts of the empire came into the hands of Daśaratha and western parts under Samprati. The last Mauryan emperor was Brihdratha who was

---

7.   D.N. Jha K.M. Shrimali(Ed.), *Prāchina Bhārata kā Itihāsa.*(Hindi), Delhi, 1986, pp. 168-169.

8.   P.V. Bapat(Ed.), *2500 Years of Buddhism*, Delhi, 1956,p.39.

9.   D.N. Jha & K.M. Shrimali(Ed.), *Op. Cit.*, p.169.

assassinated by his general Puṣyamitra Śuṅga in 185 B.C. and it was he who laid the foundation of the Śuṅga dynasty.[10]

Puṣyamitra Śuṅga who laid the foundation of the Śungan dynasty belonged to Baimbika family.[11] He ruled about thirty-six years over a maximum portion of Bihar. The *Divyāvadāna* states that Puṣyamitra was the persecutor of Buddhism and he had destroyed many Buddhist edifices constructed by Aśoka and he is placed as a staunch follower of Brāhmanism.[12] But the allegation on Puṣyamitra may not be true because we know that the Buddhist *Stūpa* at Bharhut was constructed during the Suṅgan period (C. 185-75 B.C.), even the railings of Sānchi *Stūpa* were erected during that period.[13] So, if Puṣyamitra was the persecutor of Buddhism and destroyed Buddhist edifices then how do we account for these Buddhist edifices during their rule? Puṣyamitra was succeeded by his son Agnimitra who is the hero of *Mālavikāgnimitra* of Kalidas and he was followed by his son Sujyeṣṭha or Vāsujyeṣṭha who ruled about seven years.[14] Sujyeṣṭha was succeeded by Vasumitra or Sumitra who is known to be a son of Agnimitra.[15] Vasumitra had fought with the Yavanas to rescue the sacrificial horse of his grandfather and was probably posted to guard the North-western frontier of Puṣyamitra's empire. According to Bāna's *Harsacarita* he was killed in the course of a theatrical performance by one Mitradeva.[16] After that the history of the Śuṅgan dynasty is not very much clear, however, the tenth or last ruler of this dynasty according to the *Purāṇas*, was Devabhūti or Devabhūmi who was the victim of a conspiracy engineered by his minister Vāsudeva and was killed by a slave girl who approached him in the guise of his queen. Altogether ten rulers of this dynasty ruled for a period of 112 years from B.C. 185 to 75 B.C.[17]

---

10. R.C. Majumdar (Ed.), *Op. Cit.*, pp.54,69,71,88-90.
11. H.C. Raychaudhury, *Political History of Ancient India*, Calcutta, 1950, p. 369.
12. Bela Lahiri, *Indigenous States of Northern India*, (C.200 B.C. to 320 A.D.), Calcutta, 1974, pp.33-35
13. D. Mitra, *Buddhist Monuments*, Calcutta, 1971, pp.92,97.
14. Bela Lahiri, *Op. Cit.*, p.51.
15. *Ibid.*
16. *Ibid.*, pp.51-52.
17. R.C. Majumdar (Ed.), *Op. Cit.*, pp. 97-98.

The Kāṇva dynasty was founded by Vāsudeva in 75 B.C. who was the royal minster of the last Śuṅgan ruler Devabhūti. It is said that Vāsudeva got the king assassinated and himself ascended the throne. Only four kings of this dynasty, Vāsudeva, Bhumitra, Nārāyaṇa and Susavarman are known who ruled upto circa 30 B.C.[18] According to the *Purāṇas*, the Kāṇva dynasty was uprooted by the Andhras but there is no other evidence excepting the *Purāṇas* that the Andhras ruled over Magadha or any other parts of Bihar. For a short period the Mitra dynasty is said to have ruled over Pāṭaliputra.

The political history of Bihar particularly of Magadha after the Kāṇvas and before the Guptas is not very much clear. However, the discovery of a clay seal with the legend Mokhalinam proves that the Gayā region was under the power of Maukhari chiefs, but adequate date of the record is not known. Mahārāja Trikamala probably ruled over in the same region in the year of 64 of an unspecific era. On the basis of epigraphical sources it appears that Magadha was ruled by some Mitras in the first half of the first century B.C.[19] Two of the Mitra rulers of Magadha, Indrāgnimitra and Brahmamitra are known from the Bodhgayā inscriptions,[20] while another Mitra ruler, Brihaspatimitra is known from the Hāthigumphā inscription of Khārvela.[21] But about the successors of Brihaspatimitra no adequate evidence is known. However, as is stated earlier, Gayā region came under the sway of Maukharis.[22]

A number of copper coins of the Kuṣāṇas have been discoverd from Bihar region which may indicate that this region was under the fold of the Kuṣāṇa empire.[23] But apart from the numismatic evidence we do not have any other adquate evidence, on that basis we can say that Bihar was a part of the Kuṣāṇa empire. However, it seems that the Muruṇḍas who ruled over Magadha portion of Bihar was either a viceroya of the Kuṣāṇas or an independent ruler, but one

---

18.   D.N. Jha & K.M. Shrimali (Ed.), *Op. Cit.*, p.233.

19.   B. Lahiri, *Op. Cit.*, pp. 66-68.

20.   *Ibid*.

21.   D.C. Sircar, *Select Inscriptions*, Vol.I, Calcutta, 1942, p.209.

22.   A detailed discussion on this issue have been made in B. Lahiri, *Op. Cit.*, p.69.›

23.   H.C. Raychaudhury, *Op. Cit.*, p. 473.

thing is quite clear that there was some sort of the Kuṣāṇa influence on Bihar. After the Muruṇḍas Ārya Visākhamitra ruled over the region as an independent king. It is also suggested that the Lichchavis might have ruled over Magadha before the rise of the Guptas.[24] But it is also said that before the rise of the Guptas some Koṭā family was ruling at Pāṭaliputra.[25]

The Gupta empire which was founded by Śrigupta, ruled over the maximum area of Bihar. The first important king of this dynasty was Chandragupta-I[26] who ruled from 319 to 335 A.D. He got married to a Lichchavi princess, Kumrārdevī and he is credited to have started the Gupta era in 319-20 A.D. He was succeeded by his son Samudragupta who ruled from 335-380 A.D. The reign of Samudragupta is remarkable for the series of military campaigns which he led in various parts of India. It is said that he was the opposite of Aśoka who believed in a policy of peace and non-aggression whereas Samudragupta delighted in violence and conquests. The author of the Allahabad *Praśasti*, his court poet Hariṣeṇa, wrote a glorious account of the military campaigns of his patron.[27] The places and countries conquered by Samudragupta were Ganga-Yamuna doab, eastern Himālayan states, Nepal, Assam, Bengal, Panjab, eastern Deccan and South India etc. Due to his victorious conquests he is called the Napoleon of India by V.A. Smith.[28] Samudragupta was succeeded by his son Rāmgupta who ruled for a short period, then Chandragupta II came into the throne. It was during his reign when the celebrated Chinese pilgrim Fa-Hien visited India who left valuable information regarding the Gupta period. The reign of Chandragupta II saw the high watermark of the Gupta empire. He extended the limits of the empire by marriage allianece and conquests. He established the friendly relation with the Vākāṭaka kings and married his daughter Prabhāvati with Vākāṭaka prince Rudrasena II. The Vākāṭaka ruler Rudrasena II died in 390 A.D. and was succeeded by his minor son.

24. B. Lahiri, *Op. Cit.*, pp. 70-71.
25. H.C. Raychaudhury, *Op. Cit.*, pp. 401-402.
26. R.C. Majumdar (Ed.), *The Classical Age*, Bombay, 1954, p.1.
27. *Ibid*.
28. *Ibid.*, p. 14.

his minor son. Prabhāvati took the charge of the empire as a regent. During her period the Vākāṭaka empire was very much influenced by the Guptas as Prabhāvati always sought help or advice from her father Chandragupta II. The exploits of a king called Chandra are glorified in an iron pillar inscription fixed near Qutab Minar in New Delhi. If Chandra is considered to be identical with Chandragupta II of the Gupta dynasty as it has been proved that paleographically the said inscription belong to the period of Chandragupta II. So, it appears that he established Gupta authority in North-Western India and was in possession of a good portion of Bengal.[29] Chandragupta II had adopted the title of Vikramāditya which was first used by an Ujjain ruler in 58 B.C. as a mark of victory over the Śakas. The court of Chandragupta II at Ujjain was adorned by numerous scholars including Kalidas and Amarasimha.[30] On the death of Chandragupta II the throne passed on to another person whose identity is not clearly known. However, it is generally believed that Chandragupta II was succeeded by his son Kumāragupta, born of his chief queen Dhruvadevī.[31] It is believed that Kumāragupta had maintained intact the vast empire which he had inherited from his father. He had performed an *Aśvamedha* sacrifice and assumed the title Mahendrāditya. Towards the end of his reign the peace of the empire was rudely disturbed by the invasion of Puṣyamitra as is known from the Bhitari inscription of Skandagupta.[32] Kumāragupta was succeeded by his son Skandagupta. He has to face the invasion but he effectively stemmed the march of the Huṇas. Skandagupta was not only an able army chief, but also an ideal king. He had renovated the dam of the Sudarsana lake (Girnar) which was constructed by Chandragupta Maurya. The account of Huien-Tsang indicates that Skandagupta had patronised the Nālandā University. He had also sent royal emissary to China.[33] After the death of Skandagupta, the Gupta empire began to decline and gradually different feudatories who were affiliated to the Gupta empire, declared themselves free from the yoke of the

---

29. *Ibid.,* pp. 20-22.

30. *Ibid.,* pp.302-303.

31. *Ibid.,* p.23.

32. D.C. Sircar, *Op. Cit.,* pp.312-315.

33. D.N. Jha & K.M. Shrimali (Ed.) *Op. Cit.,* p. 284.

empire. However, the Gupta empire lingered on in Bihar till the middle of the sixth century A.D.

After the break-up of the Gupta empire several new dynasties came into existence. The Magadhan region came under the power of the later Gupta rulers and the Maukharis rose into power in Kanuaj and they also held their sway in the Gayā region of Bihar.

The Maukharis who came into prominence during the sixth century A.D., were in existence from quite an earlier period.[34]The inscriptions engraved on Stone *Yupās* (sacrificial pillars) reveal that there were several Maukhari families during the third century A.D.[35] and they were widely spread over Northern India at a very early period.[36]A Maukhari family used to rule in the neighbouring area of Gaya in the sixth century A.D. Three kings of this family are known from three inscriptions discovered in the Barābar and Nāgārjuṇa Hills in the Gayā district.[37] The Maukhari dynasty declined after Grahavarman, husband of Rājyaśri (Harsa's sister), was killed by a king of Mālwā.[38] He was followed by Suva or Suvra and after that the history of Maukharis is obscure.[39]

The history of the later Guptas is similar to that of the Maukharis as they were at first the feudatory to the Imperial Guptas and taking the benifit of weakness of the last Gupta rulers, they threw out the yoke of the Gupta domain and became indepnedent.[40] The Aphsaḍ inscription of Ādityasena mentions the geneology of the early kings of this dynasty which includes the names of Kriṣṇagupta, Harshagupta, Jivitagupta etc. The later Guptas held their sway over the Magadhan region.[41]

During the reign of Harṣavardhana of Kanauj Huien-Tsang visited india who had studied at Nālandā University which was at

34. R.S. Tripathy, *History of Kanauj*, Varanasi, 1937, pp.26-27.
35. D.C. Sircar, *Op. Cit.*, pp. 92- 93.
36. R.C. Majumdar (Ed.), *Op. Cit.*, p. 67.
37. *Ibid.*
38. P.V. Bapat (Ed.), *2500 Years of Buddhism*, Delhi, 1956, p. 178.
39. R.C. Majumdar (Ed.), *Op. Cit.*, p. 71.
40. *Ibid.*, p.72.
41. D.C. Sircar, *Op. Cit.*, Vol.II, Delhi 1983, pp.44-47.

its peak at that time. Though Harṣavardhana had his capital at Kanauj but Bihar was under his domain.

After the death of Harṣa in 647 A.D., Bihar again came under the dominance of the later Guptas who revived their power and again captured the Magadhan State. The rulers of the later Gupta dynasty of Magadha ruled upto the second quarter of the eighth century A.D. The last ruler of this dynasty was Jivitagupta who was defeated either by a king of Gauḍa or king Yasovarman of Kanauj. An inscription found at Nālandā refers the name of Yaśovarman as the paramount suzerain.[42] It is also said that after the death of Harṣa, Bhaskarvarman of Assam had occupied Magadhan throne.[43]Actually the history of Bihar particularly of Magadha after the death of Harṣa and before the rise of the Pālas is not very much clear, it is a rather drab story of endemic warfare between rival dynasties.

During the reign of the Pālas, major portion of Bihar was under their domain and remaining portions were ruled by many petty dynasties.[44]

B. *BENGAL* : In modern times Bengal, which is situated between the rivers Ganga and Brahmaputra, means Bengali speaking areas of Independent India and Bangladesh. In early period it was divided into a number of small kingdoms viz., Gauḍa, Vaṅga, Sāmataṭa, Harikela, Chandradvipa, Vaṅgāla, Puṇḍravardhana, Vārendri, Rāḍhā, Tāmralipta and Suṃha. The area of these kingdoms varied in different periods. In course of time Gauḍa and Vaṅga became more powerful and the remaining states were merged with either of them.[45]

So far as the early History of Bengal is concerned, there is no mention of it in the Vedic hymns. But it does not mean that during the vedic period these areas were not inhabited by human beings. Recent researches prove the prevalence of pre-Āryan culture in these areas.[46] It seems that either the early Āryans might not have been able to penetrate the land of Bengal or the Āryana's geographical

---

42. S.Sengupta, *Buddhism in the Classical Age*, Delhi, 1985, pp. 25-26.

43. A.L. Basham, *The Wonder that was India*, Delhi, 1981,p.71.

44. H.C. Varma, (Ed.), *Madhyakālina Bhārata* (Hindi), Delhi, 1983, pp.11-13.

45. G. Senmajumdar, *Buddhism in Ancient Bengal*, Calcutta, 1983, pp. 7ff.

46. D.N. Jha & K.M. Shrimali (Ed.), *Op. Cit.*, p.107.

knowledge did not reach upto Bengal. So, Bengal was away from the Āryan pale. However, the *Aitareya Brāhmaṇa* mentions that the people who lived in large numbers beyond the frontiers of Āryandom were classed as Dasyus[47] and among such people we find the mention of the Puṇḍras having their capital city at Puṇḍranagar situated in the Bogra district of modern Bangladesh. The name of Vaṅga are mentioned in the ancient epic and the *Dharma sūtras*. In *Bodhāyan Dharmasūtra* we find the mention of the division of land known to it into three ethnic belts which were regarded with varying degrees of esteem.[48] The holiest of the three was Āryāvarta which was lying between the Himālayas and the western Vindhyas and watered by the upper Ganga and the Yamuna. The next in point of sanctity were the area of Mālwā, East and South Bihar, South Kathiāwār (Gujarat), the Deccan and the lower Indus valley. The people who were living in the area of Araṭṭas of the Punjab, the Puṇḍras of North Bengal, the Sauviras occuping parts of Southern Punjab and Sindh, the Vaṅgas of Central and Eastern Bengal, and the Kaliṅgas and adjoining area were regarded as altogether outside the pale of Vedic culture. It is interesting to note here, even a person of the holy land who lived in these areas for a short period, were required to go through expietory rites.[49] This is why these area were also known as *Vrātya* country, while the *Rāmāyana* mentions them in a list of peoples that entered into very close political relations with the high born aristocrates of Ayodhyā.[50] The *Mahābhārata* mentions that Bhīma undertook a campaign in the land we call Bengal.[51] The Jain text *Āchāraṅgasūtra* describes the land of the Lāḍhās (Rāḍhas) in West Bengal as a pathless county inhabited by a rude people who attacked on the peaceful Jain monks.[52]

During the period of the Mauryan rule, the eastern part of Bengal was under their domain as testified by the Mahāsthāna inscription which belongs to the early Mauryan period and discovered

---

47. R.C. Majumdar (Ed.), *History of Bengal*, Vol.I, Dacca, 1963, p.7.
48. *Ibid.*, 8.
49. *Ibid.*
50. *Ibid.*
51. *Ibid.*
52. *Ibid.*, p.9.

from eastern Bengal.[53] According to *Mahāvaṁsa*, Aśoka had come upto Tāmralipti to see off his son when the latter was going to Ceylon for the propagation of Buddhism.[54] The Chinese travellar Huien-Tsang mentions that he had seen the Buddhist *Stūpas* at Tāmralipti, Karṇasuvarṇa, Sāmtaṭa, eastern Bengal and Puṇḍravardhana.[55] The *Divyāvadāna* also says that Bengal was part of the Magadhan empire.[56]

After the fall of the Mauryan empire, it is not clear as to which domain Bengal came under. However, on the basis of terracotta figurines which have been discovered at Mahāsthāngarh, we may say that the city of Puṇḍravardhan continued to flourish vigorously even after the fall of the Mauryan empire. On the basis of the accounts of the Periplus and Ptolemy we can surmise that in the first two centuries of the Christian era the whole of deltaic Bengal was organised into a powerful kingdom with its capital at Gange, a great market-town on the banks of the river Ganga.[57]

Some Gold and Copper coins belonging to the period of the Kuṣāṇas have been discovered at Mahāsthāna, Mālda and Tamluk[58] etc. and a number of sculptures discovered at Chandraketugarh, Hankrail, Dinājpur etc. which have great affinity with the Kuṣāṇa art idiom[59] e.g., the broad shoulders, raised eyebrows, poses of hands etc. On the basis of these evidences it is said that the Kuṣāṇa kings might have extended their sway over the areas of Bengal. But it is a debatable matter since we do not have any further evidence regarding this. D.C. Sircar aptly says that whether the sculptures, as mentioned above, prove to the inclusion of Bengal in the Kuṣāṇa empire or the migration of sculptors who carved these images, cannot be determined

---

53. D.C. Sircar, *Select Inscriptions*, Vol.I,...,1942,pp.82-83.

54. Quoted from D.N. Jha & K.M. Shrimali(Ed.), *Op. Cit.*, p.180.

55. T. Watters, *On Yuan Chwang's Travel in India*, Vol.II, London, 1905, pp.184-185, 187-89,191.

56. Quoted from D.N. Jha & K.M. Shrimali (Ed.), *Op. Cit.*, p.180.

57. R.C. Majumdar (Ed.), *Op. Cit.*, pp. 44-45.

58. D.C. Sircar, "Eastern India and the Kushans," in *Central Asia in the Kushan period*, Vol.II (Proceedings of the International conference of the History, Archaeology, and Culture of Central Asia, in the Kushan period), Dushanbe (U.S.S.R.), 1968, p.9.

59. *Ibid.*, p.12.

with certainty, but the fact can not be dissociated from the problem of the extension of the Kuṣāṇa influence in Eastern India.[60]

After the Kuṣāṇa period some new kingdoms came into existence viz., Sāmataṭa in Eastern Bengal and Pushkaraṇā in Western Bengal which were ruled by Siṁhavarman, towards the close of the third or beginning of the fourth century A.D. and he was succeeded by his son Chandravarman who seems to have been a mighty warrior. It is said that he had extended his dominions eastwards.[61]

The rise of the Guptas in power marks the end of several independent existence of the states of Bengal with the exception of Sāmataṭa. These states were incorporated in the Gupta empire by Samudragupta. Sāmataṭa was a tributary state which acknowledged the suzerainty of the Gupta rulers with full autonomy in respect of their internal administration,[62] but gradually it was also incorporated into the Gupta empire.[63] The Guptas maintained their hold over northern Bengal till 543 A.D.

After the fall of the Gupta empire, two independent kingdoms arose in Bengal. The first kingdom comprised the southern and eastern Bengal and the second comprised the southern part of western Bengal. Six copper-plate grants discovered near Koṭalipāḍā and Burdwan refer to the names of three kings of this dynasty viz., Gopachandra, Dharmāditya and Samāchāradeva.[64] These three kings have been referred to in the period 525-575 A.D. A large number of crude and debased Gupta type coins have been discovered in different parts of eastern Bengal which prove the existence of other kings in this locality who obviously ruled later. The name of two kings, Prithuvira and Sudhayāditya have been read who had issued these coins. The Mahākuṭa inscription mentions that Chālukya king Kīrtivarman had conquered among the countries, Aṅga, Vaṅga, Kaliṅga and Magadha, and his conquest in Bengal has been placed in the last quarter of the sixth century A.D.[65] From this period, Gauḍa and Vaṅga became two

---

60. *Ibid.*

61. R.C. Majumdar (Ed.), *Op. Cit.*, p.45.

62. *Ibid.*, p.49.

63. *Ibid.*

64. R.C. Majumdar (Ed.), *The Classical Age*, Bombay, 1954, p.77.

65. R.C. Majumdar (Ed.), *History of Bengal*, Vol.I,..., p.54.

prominent political divisions of Bengal. The former roughly comprised the Northren and Western Bengal and the latter, the Southern and Eastern Bengal. It is learnt that Śilabhadra, the head of the Nālandā University at the time of Harṣa of Kanauj belonged to Samataṭa.[66]

The kingdom of Gauḍa became more prominent during the reign of Śaśāṅka who seems to be a feudatory chief of the later Guptas in the early period,[67] afterward became independent. But sometime before Śaśāṅka, one Māna dynasty had established a kingdom in the hilly region between Mednapore and Gayā regions. In course of time, this dynasty extended their sway upto Orissa. Sambhuyaśas, a ruler of this dynasty was ruling over Orissa in 580 A.D. Sambhuyaśas or his successor was defeated by Śaśāṅka who captured Daṇḍabhukti, Utkala and Koṅgoda. He had also conquered the whole of Magadha and probably even Benaras. He had also proceeded against the Maukharis and Pushyabhūtis of Thāneswar. Then Harṣavardhaṇa of Pushyabhuti family marched against Śaśāṅka and also rescued his sister Rājyaśrī whom he met in the forest. But whether Śaśāṅka ever met Harshavardhana on the battle field is not clearly known,[68] even then it is said that Harṣa had taken not only revenge over Śaśāṅka but also extended his power on other states of Northern India.[69] However, Huien-Tsang mentions that Śaśāṅka was in possession of Magadha till his death (circa 637-38 A.D.)[70] Śaśāṅka is blamed to have had cut down the Bodhi tree at Gayā,[71] this may not be ture because we find the evidence about the flourishing condition of Buddhism in the state of Śaśāṅka.[72] So, if Śaśāṅka was anti-Buddhist then why did he allow Buddhism to flourish in his own state?

After the death of Śaśāṅka, the history of Bengal is not very much clear. The political solidarity, which was brought by Śaśāṅka, was divided into a number of independent states. Just after Śaśāṅka,

---

66. R.C. Majumdar (Ed.), *The Classical Age...*, pp.77-78.

67. *Ibid.*

68. *Ibid.*, pp.79-80.

69. D.N. Jha & K.M. Shrimali(Ed.), *Op. Cit.*, p.287.

70. T. Watters, *Op. Cit.*, Vol.II, p.92; R.C. Majumdar (Ed.), *Op. Cit.*, p.80.

71. T. Watters, *Op. Cit.*, p.115.

72. R.C. Majumdar (Ed.), *Op.Cit.*, p.80.

the political condition of Gauḍa was reduced to mutual mistrust which led to the civil war and in course of which one king ruled for a week another for a month and then a republic was established and after that Śaśāṅka's son captured the throne, but he ruled only for a few months. It is said that taking the benifit of this anarchy and confusion, Harṣa and Bhāskaravarman invaded on the Gauḍa kingdom and each of them incorporated some parts of Bengal in their empire for sometime.[73] Not after a great interval of time another powerful king came into the throne of Gauḍa whose name was Jayanāga. He was contemporary to Bhāskarvarman of Assam; he issued a land- grant from Karṇasuvarna. The successor of Jayanāga is not known, however, it is generally believed that the kingdom of Gauḍa passed into the hands of later Gupta rulers.[74]

Of Vaṅga, Huien-Tsang says that some Brāhmaṇa kings ruled over this territory in the first half of the seventh century A.D. Śilabhadra, once the head of Nālandā University, was a member of this family.[75]

It is said that in the latter half of seventh century A.D., some kings of a Khaḍga dynasty were ruling in East Bengal and over a considerable part of Southern and Central Bengal.[76]

In the first half of the eighth century A.D. Bengal witnessed a series of foreign invasions. A king of the Śaila dynasty had conquered the portion of Northern Bengal, but except this nothing is known about their activities in Bengal. It is said that originally they ruled over the Himālyan region and later on moved to east and south.

Later on, sometime between 725 and 735 A.D., West and East Bengal was conquered by Yaśovarman but his conquest was short-lived and he had to acknowledge the suzerainty of Lalitāditya of Kashmir as Kalhaṇ mentions that Yaśovarman of Kanauj was defeated by Lalitāditya.[77] Lalitāditya had also annexed Gauḍa into his empire, but later on, Gauḍa regained its independence.

---

73. *Ibid.*, p.142.
74. R.G. Basak, *History of North-Eastern India*, Calcutta, 1934, p. 128.
75. T. Watters, *Op. Cit.*, p.109.
76. R.C. majumdar (Ed.), *Op. Cit.*, p.143.
77. Stein (Tr.), *Rājtaraṅginī*, Vol.I, IV, Verses, 132,139.

Gradually, the whole of Northern and Western Bengal was split into a number of independent states and before the rise of the Pālas, the condition of Bengal was a period of anarchy and confusion caused by political disintegration. About this condition of Bengal a contemporary record rightly describes by the well-known term '*Mātsyanyāya*' (rule of the fish) which meant that a state of anarchy in which might alone is right, as in a pond where the stronger fishes devour the weaker ones.[78]

As stated above, before the rise of the Pālas the condition of Bengal was not peaceful and this led to a bloodless revolution by which a local chief named Gopāla was elected by the people as their ruler.[79] This is however, to be taken not a strictly democratic or constitutional sense which was perhaps not possible in those days. Anyway, in this way Gopāla led the foundation of the Pāla empire. Though the details of Gopāla's career are not known, he introduced peace in the kingdom and laid the foundation of the future greatness of his family.[80] Tāranāth[81] mentions that Gopala had built the celebrated monastery at Odantapuri and ruled for the period of forty-five years. Gopāla was succeeded by his son Dharmapāl who was an energetic personality, and the task of interval consolidation having already been accomplished by his father, he found himself in a position to undertake foreign expeditions. Among his achievements, the defeat of Indrāyudha of Kanauj was most notable. He is said to have founded the famous Buddhist monastery at Vikramaśilā (modern Antichak in Bhagalpur). The monastery bears eloquent testimony to his liberality as well as to that of others donors.[82]

The Pālas ruled over Bengal upto the first quarter of twelfth century A.D. then the Senas came into the scene.

---

78. R.C. Majumdar (Ed.), *Op Cit.*, p.144.
79. R.C. Majumdar (Ed.), *The Age of Imperial Kanauj*, Bombay, 1955, p.44.
80. *Ibid.*, pp.44-45.
81. Taranath's *History of Buddhism in India*, (Ed.), D.P. Chattopadhyaya, Calcutta, 1980, pp.257ff.
82. R.C. Majumdar (Ed.), *Op. Cit.*, pp.45-50.

C. *ORISSA* : In modern times Orissa is one of the states of Indian Union, located in eastern coastal area. Orissa yielded a good number of Buddhist images in early medieval period. Orissa first came into prominence during the Kaliṅga war of Aśoka. A strong state was founded in that area only in the first century B.C. Its mighty ruler Khāravela is said to have hoisted his flag of victory as far as Magadha.

Kaliṅga and Ukhala or Ukkala are the two ancient *Janapadas* comprising the area of the modern state of Orissa. However, the *Aṅguttara Nikāya* does not list them among the sixteen *Mahājanapadas*. But it is mentioned in the extended list of the *Niddesa*.[83] We also find numerous references of Kaliṅga with its capital, Dantapura in the Pāli *Jātakas*[84] etc. The Ceylonese chronicles mention that there was political relation between the people of Kaliṅga and Vaṅga from very early times.[85] Huien- Tsang says that Kaliṅga country was above 5000 *li*.[86]

The *Vinaya Piṭaka* mentions that the two merchants Tapassu and Bhallika were on their way from Ukkala to Majjhimadeśa when they offered cake and honey to the Buddha at *Rājāyatanamula* tree in Uruvela,[87] and became the first lay-converts of the Buddha.

During the reign of the Nandas, Kaliṅga was probably under their dominions as indicated by the Hāthigumphā inscription of Khāravela.[88] But B.M. Barua opined that Kaliṅga remained unconquered till the seventh year of Aśoka's reign.[89]

The *Mahāvastu* placed Utakala in the Uttarāpatha[90] and the *Mahābhārata* mentions Utakala several tims in the list of tribes.[91]

During the Mauryan period Orissa was properly annexed by Aśoka into his empire just after the Kaliṅga war. But after the fall

---

83. Cullaniddesa, ii,37, (PTS).

84. *Jātaka*, ii.367f; *Mahāvastu*, iii,361.

85. *Mahāvaṁsa*, VI,I; Dipavaṁsa, IX,2ff:

86. T. Watters, *Op. Cit.*, Vol. II, p.198.

87. *Vinaya*, 1.4 (PTS).

88. H.C. Raychaudhury, *Op. Cit.*, p.229-230.

89. B.M. Barua, "Hathigumpha Inscription of Kharavela", *IHQ*, Vol.XIV, 1938, p.276n.

90. *Mahāvastu*, iii.303.

91. *Bhiṣmaparva*, ix.365; *Droṇaparva*, iv, 122.

of the Mauryan empire the history of Orissa is obscure. It is not very clearly known when Orissa exactly threw off the yoke of the Magadha and came under the power of the Chedis. The Hāthigumphā inscription and some other sources indicate that Mahā-Meghavāhana was the founder of the Chedi dynasty.[92]

The Mañchapuri cave inscription of the chief queen of Khāravela mentions that the cave was excavated for the Jain monks and the other inscriptions from the same cave indicate that Vakradeva was a second generation of the royal family and may be the father of Khāravela.[93] Khāravela is considered as one of the remarkable kings of early India. He assumed the title of Kaliṅgādhipati and claimed the status of a universal ruler. Khāravela was a dvout Jain and was called *Bhikshu-rājā*, i.e. the monk- king,[94] but he devoted considerable time in invasions on the neighbouring areas. It is said that he had invaded a city to the south of Sātakarṇi's kingdom. The states which were either invaded or destroyed by Khāravela's army are, Berar, Gorathagiri, Rājagriha, etc. A passage of the Hāthigumphā inscription indicates that Khāravela had killed a king named Gorathagiri.[95] As a devout Jaina he had excavated a number of caves and a monastery dedicated to the cause of Jainism.[96]

After Khāravela, the history of Mahā-Meghavāhana (Chedi) dynasty is obscure. However, it seems that not long after Khāravela, the country was split up into a number of smaller principalities.[97]

It is hardly known to us about the political history Orissa of the Gupta period. However, it is said that Orissa was directly administered by the Gupta emperors. The trace of the Gupta influence in the Gupta year is the Ganjam inscription of Mādhavavarman II of the Śailodbhava dynasty of Kongoḍa, a feudatory of the Gauḍa king Śaśāṅka.[98] A copper-plate grant records the names of Prithvivigraha

92. R.C. Majumdar (Ed.), *The Age of Imperial Unity*…,p.212.

93. D.C. Sircar, *Select inscriptions*, Vol.I.,1942., pp.213- 4., Cf. R.C. Majumdar (Ed.), Op. Cit., pp.212-213

94. R.C. Majumdar (Ed.), *Op. Cit.*, p.213.

95. D.C. Sircar, *Op.Cit.*, p. 208.

96. R.C. Majumdar (Ed.), *Op. Cit.*, pp. 214-215.

97. D.C. *Sircar*, 'Gupta Rule in Orissa', *IHQ, Vol.XXVI, No.1, 1950, p.75.*

98. *Ibid.*, p.76.

Bhaṭṭāraka and Mahārājā Dharmarājā as governors of the Gupta empire in Orissa.[99]

In the last quarter of the sixth century A.D. Māna and Śailodbhava families were ruling respectively in the Northern and Southern parts of Orissa. Most probably they rose to power on the ruins of the Gupta empire. The kingdom of Māna family very soon came into the possession of Śasāṅka, but after the death of Śaśāṅka, Orissa gained independence and the Śailodbhavas who came into the power extended their sway over the neighbouring areas.[100] Though it is said that Harṣavardhana had annexed some portion of Orissa into his empire in course of his eastern campaign.[101] However, Śailodbhavas seem to have ruled over Orissa upto the middle of eighth century A.D. They were most probably ousted by the Karas.

The Karas who were also called Bhauma, were contemporary to the Pālas. The Bhaumas were so called because they claimed descent from Bhumi or Earth and Karas derived from names of all the kings of the family as ended in Karas.

The reign of the Bhauma-Karas is considered to be a glorious age in the history of Orissa. Buddhism got the affectionate patronage of the first group of Bhauma-Karas, many of whom were devout Buddhists.[102] The Bhaumakaras were ousted by Somavaṁsis in circa later tenth century A.D.[103]

8.2 : Buddhism in Eastern India :

A. *BIHAR* : Bihar is considered as the sanctum sanctoram of early Buddhism. Though the Buddha had delivered his first sermon at Ṛsipattana (Sārnāth), but elements of Buddhism might have been preached for the first time in Bihar (Bodhgayā) by the Buddha himself to Tapassu and Bhallika who were on their way to Bodhgayā when they met the Buddha just after His enlightenment and became His

---

99. *Ibid.*, p.77.

100. R.C. Majumdar (Ed.), *Op. Cit.*, pp.93-95.

101. R.S. Tripathy, *Op.Cit.*, p.106.

102. B. Mishra, *Orissa under the Bhauma Kings*, Calcutta, 1934, pp. 40-51.

103. R.C. Majumdar (Ed.), *Op. Cit.*, p.68.

converts. The different parts of Bihar were frequently visited by the Buddha. He had spent several *Vassāvāsas* in different places of Bihar. In Bihar the first three Buddhist councils were held at Rajgriha, Vaishāli and Pāṭaliputra, respectively.[104]

The reign of Aśoka is considered as the glorious period for Buddhism in Bihar. Aśoka who himself was a lay-follower of Buddhism imparted royal patronage to Buddhism. During his reign the third Buddhist council was held at Pāṭaliputra under the presidentship of Mogaliputta Tissa and after the completion of this council Buddhist emissaries were sent to different parts of the country and abroad.[105]

The Mauryas were followed by the Śuṅgas and the Kāṇvas, respectively who supported Brāhmanism however the construction of the railing at Bodhgayā during the Śuṅgan period proves the tolerance of the monarchs of this dynasty towards Buddhism.[106]

The Gupta period which is considered as the golden Age in the history of India witnessed an immense development in art and culture. During this period, though Buddhism was not a state religion, it was in a flourishing condition. Fa-Hien (5th century A.D.) found Buddhist monasteries in flourishing condition in Vaishāli, Pāṭaliputra and Rājagriha etc. During this period Nālandā began to flourish as a centre of learning. The inscription of Mahārājā Trikamala, probably a feudatory of the Guptas, records that two monks, both the teachers of *Vinaya*, built one Simharatha for dedicating an image of Bodhisattva with the help of a lady lay- devotee and expert of the holy text. Two other inscriptions of the Ceylonese monk, Mahānāma dated to the 6th century A.D. mentions about the erection of a *Maṇḍapa* for the Buddha figure and the gift of an image of the Buddha within the area of Bodhgayā.[107]

Harṣa who was considered as one of the last great patrons of Buddhism, contirbuted a lot for the propagation of Buddhism, particulary in Bihar. He gave his full support to the Nālandā University. He remitted the revenue of about a hundred villages as

---

104. P.V. Bapat (Ed.), *2500 Years of Buddhism*, Delhi, 1976(reprint), pp.190,31-38.
105. R.C. Majumdar (Ed.), *Op. Cit.*, p.84.
106. B.N. Chaudhury, *Op.Cit.*, p.23.
107. *ASI Ann. Re.* 1922-23, p.169.

an endowment to the convent and two hundred householders of those villages regularly contributed to the requirements of the University. The monks of Nālandā were held in great esteem by Harṣa, this is why at his invitation a considerable number of monks from Nālandā were present in the Kanauj Assembly.[108] Huien-Tsang witnessed the flourishing state of Buddhism in Vaishālī, Nālandā, Pāṭaliputra etc.[109]

When Buddhism began to decline in the seventh century A.D. onwards in some parts of Northern India, it was still kept alive in Bihar. Rise of the Pālas in Bengal and Bihar under Gopāla proved a further boost in Buddhism. The Pālas were great patrons of Buddhism. Vikramśilā, Odantpuri (Bihar) monasteries were established during this period. This period witnessed the decline of the Hīnayāna and the Mahāyāna Buddhism and emergence of the Tāntric Buddhism in which original ethical and philosophical principles were super-imposed by esoteric yogic system combined with diverse forms of worship and rituals. This is also the period when Buddhism and Brāhmanism were coming closer. Archaeological finds suggest that Buddhism survived in Bihar till the 14th century A.D.[110]

B. *BENGAL :* When exactly Buddhism penetrated the land of Bengal can not be determined with certainty. However, the *Divyāvadāna*[111] states that Sumāgadhā, a daughter of Anāthapindaka was married to a youth of Puṇḍravardhana. The members of his family were Jain by faith. Once, Sumāgadhā, with a view to convert them into Buddhism, meditated on the Buddha and then the Buddha along with his followers reached Puṇḍra through air and preached His doctrine there. But this legend cannot be considered as a historical fact in the absence of other evidence. The Chinese pilgrim Huien-Tsang mentions that he had seen an Aśokan tope somewhere in Puṇḍravardhana where the Buddha had preached sermons.[112] It is also said that Buddhism might have obtained a footing in Northern

---

108. S. Sengupta, *Op. Cit.*, p.8.

109. T. Watters, *Op.Cit.*, Vol I,p.II.

110. B.N. Chaudhury, *Op. Cit.*, pp.26-29; *ASI. Report*, 1871-72, Vol.III, pp.126-128.

111. *Avadāna-Kalpalatā* (Sanskrit Sāhitya Parishad Ed.), p.94.

112. T. Watters, *Op. Cit.*, Vol.II, pp.184-185.

Bengal even before the Aśoka's time.[113] The Mahāsthāna inscription records the name of *Saḍvagiya* which may refers to the *Chavaggiya Bhikkhus*, the group of six *Bhikkhus* headed by Devadatta and forming an anti-party of the organisation of the Buddha.[114] The inscription is believed to be of pre-Aśokan era. But scholars are not unarimous regarding the resemblance between the *Saḍvagiya* and the *Chavaggiya*. The two votive inscriptions at Sānchi belonging to the second century B.C. records the gifts of two inhabitants named Dhamata i.e. Dharmadatta and 'Isinadana' i.e. Risinanda of Punavadana which undoubtedly stands for Puṇḍravardhana.[115] A terracotta tablet which probably belongs to the Śuṅga period found at Tāmralipta (Tamluk), is supposed to contain a scene from a *Jātaka*.[116] The Nāgārjunakoṇḍa inscription which may be dated in the second or third century A.D. refers to Vaṅga as an important centre of Buddhism. It states that Vaṅga was one of the many well-known countries which were converted to Buddhism by Ceylonese monks.[117]

On the basis of above mentioned sources we may say that Buddhism might have appeared in Bengal atleast in the early years of the christian era.

Buddhism was in prosperous condition in Bengal during the Gupta period and onwards. The Gunaighar Copper-plate of the reign of the Gupta family, dated 506-07 A.D., records the grant of land by the king in favour of the Buddhist Avaivartika Saṅgha of the Mahāyāna tradition.[118] The grant also refers to two other Buddhist monasteries in the neigbhouring area, one of which is called *Rāja-Vihāra* or the royal monastery. The condition of Buddhism in the 7th century Bengal is reflected in an inscription on a Copper-plate discovered at Kailan in the Tippera district of Bengal which mentions the donation of an official of a king named Śrīdharaṇa Rāta to the Buddhist *Triruntu* etc.[119] The rulers of the Khaḍga dynasty, who ruled over Bengal

---

113. R.C. Majumdar (Ed.), *History of Bengal...*, pp.411-412.

114. D.C. Sircar, *Select Inscriptions*, Vol.I, pp.82-83.

115. *EI*, Vol.II, p.108.

116. B.N. Chaudhury, *Op. Cit.*, p.197.

117. S. Sengupta, *Op. Cit.*, p.28; EI, Vol.XX, p.23.

118. R.C. Majumdar (Ed.), *History of Bengal...*, p.413.

119. S. Sengupta, *Op. Cit.*, p.29.

during the early part of the 8th century A.D., were devout Buddhists.[120]

The Pālas were staunch followers of Buddhism and called themselves '*Parama-Saugata*'. It was during this period that the Mahāyāna Buddhism under their patronage became a powerful force and exercised predominant influence not only in Bengal, but in other parts of India also. The Pāla period also witnessed the full development of Tantric Buddhism and also an intermixture of Buddhism and Brāhmanism which is revealed in inscriptions, Buddhist texts and sculptures of the period, collected from different parts of Bengal.[121]

C. *ORISSA* : It is believed by some of the scholars that Buddhism was introduced into Orissa by the two merchants, Tapassu and Bhallika. As it is well known that the two merchants Tapassu and Bhallika belonging to the country of Ukkala,[122] while passing through the forest where the Buddha was enjoying the bliss of emancipation under the *Rājāyatana* tree in the eighth week after his Enlightenment, met Him and reverentially bowed down before the Master and offered some rice cakes and honey and became his first lay disciples. The Buddha is said to have become extremely pleased at their offer and gave them eight handfuls of His hair. It is said that when these two merchants returned back to their home state, there they enshrined them in a *Stūpa*.[123] But no such type of *Stūpa* so far has been discovered either in Orissa or in Burma. Moreover, according to other scholar like Lokesh Chandra, Tapassu and Bhallika hailed from Bālhīka, the ancient Bālkh.[124]

It seems that Buddhism might have been systematically introduced in Orissa during Aśokan period if not earlier. Tissa, the brother of Aśoka, later known as Ekavihāriya, spent his retirement in the Kaliṅga country and Aśoka had built there a monastery for

---

120. *Ibid.*, pp.7,30.

121. *Ibid.*, pp.29- 30.

122. The Burmese tradition locate Ukkala in Burma and consider them Burmese, Cf., C. Eliot, *Hinduism and Buddhism*, Vol.III, London, 1968 (reprint), p. 50.; G.P.Malalasekera, *DPPN*, Vol.I p.991, Vol.II,p.367

123. N.K. Sahu, *Buddhism in Orissa*, Utkal University, 1958. pp.9-10

124. Cf. Lokesh Chandra, *Three Iranian words in the Iranian Traditions; Encyclopaedia of Buddhism*, Vol.II, pp.685-87

him, known as the Bhojakagiri *Vihāra* which became the centre of Theravādins school.[125] Thus, under Aśoka, Kaliṅga turned to be an important centre of Buddhism.

After the fall of Aśokan empire and bofore the rise of the Bhaumakaras, the history of Buddhism in Orissa is obscure, however, there are some stray references for the existence of Buddhism of that period. It is said that Mahādeva, the leader of the Mahāsaṅghikas, propagated the faith in the Kaliṅga region.[126] The Nāgārjunakoṇda inscription of 3rd century A.D. also mentions about the activities of early Buddhism in Kaliṅga.[127] However, the Chedis, who established their authority in Kaliṅga by the middle of the first century B.C., favoured Jainism, but, Buddhism even without royal patronage continued to flourish in Orissa as indicated by the discovery of four railing pieces, probably of a *Stūpa* from the adjoining area of the Bhāskareśvara temple.[128] Huien-Tsang mentions that he had seen many Buddhist monasteries and numerous Buddhists in the Wu- Tu (Oḍra or Orissa) country.[129] The Buddhist monasteries situated in the Jajpur hills known as Udayagiri, Lalitagiri, and Ratnagiri have yeilded a large number of Buddhist sculptures.[130] Apart from this, many votive *Stūpas* and architectural and sculptural fragments have been discovered from the site of Ratnagiri.[131] The reign of the Bhauma-Kara rulers in Orissa is considered to be the golden age in the history of Buddhism in Orissa. The early rulers of this dynasty gave their affectionate patronage to Buddhism. They constructed Buddhist monasteries, *Stūpas* etc. In this period, Kapāri, Khaḍipaḍa, Solampur etc. is considered as the important centres of Buddhist art, Guhadevapālaka (modern Jajpur), the capital of Bhauma-karas was a famous seat of

---

125. N.K. Sahu, *Op. Cit.*, p.23; *DPPN*, Vol.I, Delhi, 1983, p.585.

126. S.S. Tripathy, *Buddhism and other Religious Cults of South-East India*, Delhi 1988, p.64; N.K. Sahu, *Op. Cit.*, pp.45-46.

127. D.C. Sircar, *Op. Cit.*, pp.224-225.

128. D.K. Gangoly, "Buddhism and Jainism in Orissa during the Pre-Christian Centuries", (*Proceedings of International Seminar on Buddhism and Jainism, Orissa,1978*) p.242.

129. T. Watters, *Op. Cit.*, Vol.II, p.195.

130. S. Sengupta, *Buddhism in the Classical Age*, Delhi, 1985,p.31.

131. *Ibid.*

Buddhist learning. The Buddhist scholars from Orissa were also sent to China during this period.[132]

R.P. Chanda says that "according to the Neulpur grant of Mahārāja Subhakaradeva, a line of Buddhist kings reigned in Northern Toṣāli (roughly Northern Orissa) in the eighth century A.D. The first king of the dynasty Kṣemankaradeva called himself *Paramatathāgata*,[133] Sivakaradeva, the son of Kṣemankara is called *Paramatathāgata*, the devout worshipper of *Tathāgata* (Buddha) and his son and successor Mahārājā Subhakaradeva called himself *Paramasaugata*.[134]

A number of Buddhist images discovered in the hill tracts of the Cuttack district bear inscriptions, in the characters they are similar to those used in the Copper-plate grant of Subhakaradeva, so such type of images may be assigned to the period of Subhakaradeva.

Thus, Bhaumakara rulers not only participated in the furtherance of Buddhism in Orissa, but also made contribution to the construction of Buddhist monasteries. They also granted villages for the maintenance of Buddhist *Vihāras* etc.[135]

In this way we find that Buddhism acquired a stronghold in Orissa particularly under the Bhaumakaras. The Bhaumakaras were tolerant towards non-Buddhistic religion and made donations to both Buddhism and Brāhmanism. The outcome of their tolerant religious policy was that both Buddhism and Brāhmanism were coming in closer relations with each other.[136] It was also Orissa where Buddhism had maintained their hold even in the later period when the faith was gradually disappearing from the rest of the country mainly due to the Muslim invasions.[137]

---

132. D.C. Ahir, "Buddhism and Orissa", (*Proceedings of International Seminar on Buddhism and Jainisim*, Orissa, 1978) p.152.

133. *EP.*, Vol.XV, pp.363-64; T. Watters, *Op. Cit.*, Vol.II, p.196.

134. B.N. Chaudhury, *Op. Cit.*, p.201.

135. D. Mitra, *Op. Cit.*, p.223.

136. S. Sengupta, *Op. Cit.*, p.32.

137. *Ibid.*, p.31.

8.3 : Eastern school of Buddhist art:

A. *BIHAR* : The history of Buddhist art in Bihar can be traced back from the days of the Mauryas. The monolithic Aśokan pillars surmounted by animal figures are mythologically supposed to be related with the Buddha. The Aśokan pillars found at Lauriya-Nandangarh, Basārh-Bakhirā (Vaishāli) and Rāmpurvā are crowned with single or multiple figures of lion which may symbolise the Buddha, as the Master was also described as Sākya-Simha (lion of the Sākya race) in Buddhist texts.[138] The aesthetics of the Mauryan columns occupy a proud position by reason of their free and significant artistic form in space, the rhythmic and balanced proportion of their constituent elements, the unitary and integrated effect of the whole.[139]

The bas-reliefs of Bodhgayā are illuminating commentaries of Indian life and attitude towards life as depicted in Buddhism. The reliefs seem to be the same as the *Charana-chitras* translated into stone. The methods of arrangements of various scenes depicted on the upright pillars in more or less continuous narration which invariably suggest similar arrangements of scenes in 'Charan-Chitra' or *Paṭa- Chitra'* of later tradition.[140]

During the days of the Kuṣāṇas, Mathura and Gandhāra became the dominant schools of art. Many Buddhist sculptures from Mathura were sent to various places of Bihar viz., Rājgir, Kumrāhār, Bodhāgayā etc.[141] However, Bihar continued to be a centre of Buddhist art and kept alive its existence for a long period. The two-sided relief, each side depicting a *Vṛkṣadevatā* carved from a buff sandstone - not the red sandstone of Mathura, stylistically resemble with the Mathura's ideals. A grey sandstone Bodhisattva torso from Kumrāhār also shows clear affinity with the style of early Kuṣāṇa images of Mathura school.[142]

138. K.K. Dasgupta, "Aśokan Art - why and how for Buddhist," *Proceedings of the Indian History congress*, Varanasi, 1969, pp.56-57.

139. R.C. Majumdar (Ed.), *The Age of Imperial Unity*, Bombay, 1953, p.509.

140. *Ibid.*, pp.510-511.

141. F.M. Asher, *The Art of Eastern India*, 300-800, Oxford University Press, 1980, p.10.

142. *Ibid.*

During the days of the Guptas, Sārnāth became prominent school of art. In this period the style and ideals of Sārnāth school also influenced the images of Bihar. The standing bronze image of the Buddha from Sultanganj, near Bhagalpur and the huge metal image of the Buddha from Nālandā reveals that sculptors of these images inherited the ideals of Sārnāth school and endowed them with delicacy, spiritual refinement, emotion and sensuous appeal in their appearance.[143]

During the reign of the Pālas, the Mahāyāna- Vajrayāna form of Buddhism became prominent consequently, many Tāntrik images came into existence, however, the images of the Buddha did not loose its importance. Many images of the Buddha have been discovered from different parts of Bihar, belonging to this period. Here we find many *Sthanaka* and *Āsana* types of the Budha figures which illustrates different incidents of the life of the Master e.g. the eight miracles, including the depiction of the main central seated or standing image, which were connected with the four principal incidents, birth, enlightenment, first sermon and great decease and the four others, taming of Nalagiri, the wild elephant which was set upon him by his cousin, Devadatta at Rājagriha, his descent at Sāṅkāsya from the *Trāyastriṁsa* heaven after preaching the Dhamma to His departed mother, the acceptance of the honey offered by a monkey at Vaishāli and lastly, the great miracle in which the Buddha simultaneously multiplied himself in the presence of king Prasenjit of Kośala and several others who were present at that moment. Actually king Prasenjit saw the Buddha at everyside wherever he moved. So as far as the style of the images are concerned in this period, two styles were apparent, one the final formalized outcome of post-Gupta developments, the other a highly refined new style which is known as Eastern school style or ideals.[144] During the period of 700-800- A.D. we often find the well-known Buddhist creed mainly inscribed on the pedestal of the images_*"Ye dharmā hetu prabhavā hetuṁ teshāṁ Tathāgato hy-avadat teshāṁ cha yo nirodha evaṁ-vādī Mahā Śramaṇaḥ"* (Tathāgata(i.e. the Buddha) has revealed the cause of these phenomenon which

---

143. R.C.Majumdar (Ed.), *The Classical Age*....pp. 523-524.
144. R.C.Majumdar(Ed.), *The Age of Imperial Kanauj*, Bombay, 1955, pp. 274-275.

proceed from a cause as well as (the means of) their prevention. So says the great Monk).

B. *BENGAL* : The history of Buddist art in Bengal began with the Gupta rule if not earlier. But in this period, the terracotta art was well developed in Bengal particularly during the Śuṅga period. A Yakṣi image discovered at Tāmralipti belonging to this period, is considered as a superb work of art.[145] A burnt clay figure of a female image of the Mauryan or Śuṅgan period was discovered at Mahāsthāna[146] A few sculptures discovered in North Bengal, are identified as products of the Kuṣāṇa period. Among them there is a Viṣṇu figure from Hankrail in modern Mālda district and two Surya images, one from Kumarpur and the other from Niyāmatpur both in the Rajasahi district. All these figures are in low flat relief and are influenced by Mathura school.[147]

The sculptures of the Gupta period from Bengal mainly exhibit the well known characteristics of Sārnāth school of art combined with the emotionalism of its eastern version in more or less degree. In the 6th century we find the culmination of the classical Gupta tradition in India. A few specimens of this phase of Indian plastic art and its derivation are to be found at Pahārpur.[148]

The art of the Pāla period is considerably more widespread and more abundant than the art of any earlier century in Bengal. Most of the sculptures are carved on the Stele. During this period undoubtedly, under the impact of Tāntrik Buddhism, a large number of the images of Buddhist Pantheon came into light. The male and female divinities have the same features of full fleshly, graceful and roundness. The images of the Buddha of this period are in different poses and generally two styles were prominent, one the final formalized outcome of the post-Gupta evolution and the other a highly refined new style.[149]

---

145. F.M.Asher, *Op. Cit.*, p. 10.

146. R.C.Majumdar(Ed.), *The History of Bengal*, Vol.I....p. 521.

147. *Ibid*.

148. *Ibid.*, pp. 522-525.

149. R.C.Majumdar(Ed.), *The Age of Imperial Kanauj...*,pp. 274-275.

C. *ORISSA* : The history of Buddhist art in Orissa may be traced back from the Mauryan period. The Aśokan monuments have been mostly found in Dhauli- Bhuvaneśvara region which was the famous centre of Buddhism in the later half of the 3rd century B.C. The remains of *Stūpas*, monasteries , *Caityas* and pillars have been also noticed in this region.[150] Huien-Tsang, who visited Orissa in the first half of the 7th century A.D. mentions about Aśokan *Stūpas* situated in the south of the capital of Kaliṅga (near Toṣāli).[151] Dhauli and Jaugaḍa, from where Aśoka's rock edicts have been discovered, are surrounded by the debris of Buddhist antiquities.[152]

The so-called large liṅgam in the Bhāskareśvara temple is a subject of discussion among scholars. B.M.Barua is of the opinion that "the stump of Aśoka's monolith which is being worshipped as phallic emblem in the Bhāskareśvara temple may still bear a copy of minor pillar edict."[153] Actually the liṅgam does not possess the features of an ordinary Śiva liṅgam and seems that it was originally a huge pillar which was in subsequent period converted into a liṅgam.[154] It is notable here that the so called liṅgam is as long as 9ft. in height and 12ft. in circumference at the base, whereas the Śakti is only about 20ft. in outer circumference; so obviously the lingam is nothing but the stump of a monolithic column.

The discovery of the railing of *Stūpa*, a few yards to the north of the Bhāskareśvara temple indicates the existence of Buddhism in the post-Aśokan period in Orissa. The railing is decorated with the male and female figures which are stylistically similar to the reliefs of Bharhut, Sānchi and Bodhgayā.[155]

A large number of Buddhist sculptures have been discovered in different places of Orissa viz., Udayagiri, Ratnagiri, Lalitagiri, Solampur, Khaḍipaḍa etc. which are aesthetically very refined.

---

150. N.K.Sahu, *Op. Cit.*, p. 28.
151. T.Watters, *Op. Cit.*, pp. 193ff.
152. N.K.Sahu, *Op. Cit.*, p. 29.
153. B.M.Barua, *Aśoka and His Inscriptions*, part II, Calcutta, 1946, p. 3.
154. N.K.Sahu,*Op. Cit.*, p. 29.
155. *Ibid.*, pp. 32-33.

Thus, we may conclude that the eastern school of Buddhist art started with the Mauryan period and continued to flourish at least upto circa 1200 A.D. which produced a large number of Buddhist images which are aesthetically well refined.

8.4 : Iconography of the Buddha images

A. *BIHAR* : Stylistically the images of the Buddha from Bihar concerning the period under review can be divided into three sections:

SECTION-1: This section deals with the images belonging to the early Gupta period. The images of this period mainly derived their inspiration from the early Kuṣāṇa sculptures of Mathura school, however, the images possessed some special pecularities which are as follows:

1. Though the face is especially characteristic of the Gupta sculptures, but unlike the Kuṣāṇa images from Mathura, the cheeks are full and smooth, lips are thick and the eyes are contemplative which clearly distinguish from any image belonging to the Kuṣāṇa period.

2. The changes also can be noticed in the eyebrows.

3. Though the drapery is almost similar to the Kuṣāṇa images from Mathura but there is distinction which is perceivable in the folds of drapery, in the previous one it is more raised and prominent whereas in the latter it is more flat.

4. There are auspicious lines round the neck.

Some of the specimens which bear above mentioned features are as follows:

1 A seated image of the Buddha from Bodhgayā can be taken into the account as one of the best examples of this section.[156] The image was found near a small ruined temple about 20ft. south of the railing enclosing the Mahābodhi temple at Bodhgayā. The image wears transparent drapery which covers the left shoulder only. The position of the left fist (now broken) placed on the left knee and the right hand which is completely missing, might have been raised in *Abhaya*

---

156. F.M.Asher, *Op. Cit.*, pl.11.

mudrā. There are three round lines on the neck (*Trivali*). The earlobes are elongated, hair is arranged in curly manner, *Uṣniṣa* is prominent and the eyes are large. There is an inscription on the image which records the dedication of a Bodhisattva by one Siharatha on the fifth day of the third months of the year 64 during the reign of some Mahārājā Trikamala.

The image was previously considered as an early Gupta work imported from Mathura, now it is considered that the image was probably locally made. The stone often assumed to indicate a Mathura provenance is reddish, not the mottled Sikri sandstone so commonly used for sculptures in Mathura. Though the stone of this colour is not available near Bodhgayā but it occurs as near as Sasaram in the adjoining areas of Shahabad district. So, we may say that the image was made somewhere in the Magadhan region.

2. The next representative of this section is a sandstone Buddha head discovered at Kumrāhār.[157] The execution of the image resembles with the late fifth century figure of Sārnāth sculptures. The overall face is a little-bit rubbed, the nose is partially broken, the eyes are half-closed as in meditation, the lips are thick and full, the earlobes are elongated and the hair is arranged in curly manner in which *Uṣniṣa* is clearly visible.

3. A life-size standing Buddha image from Bodhgayā, now in the possession of Bodhgayā Mahant is the next representative of this section.[158] This black stone Buddha image shows little elegance that imbued much of the art of eastern India of the end of 5th century A.D. Stylistically the image is influenced by Sārnāth school. The posture is rather broad and massive which recalls the images from the Mathura school datable to the period first half of the 5th century A.D. The image has got a large head with visible *Uṣniṣa*. The curly hairs are very thick set which looks like a cap. The eyes are half closed, eyebrows are prominently raised and the lips are thick which resemble with the Buddha image from Maṅkuār. The edge of the halo (*Prabhāmaṇḍala*) is scalloped which reminds the halo of the Kuṣāṇa period. The noticeable thing is that here the halo is replaced by large almond shaped back piece.

---

157. *Ibid.*, pl.25.
158. *Ibid.*, pl. 24.

SECTION-2: The images of the Buddha belonging to the period circa 550-700 A.D. show some short of gowth in style. Now we find a contrast between the Sārnāth inspired forms of the Rājgir-Nālandā area and the Magadhan style of Gayā area. At Nālandā, the stucco sculptures of the great *Stūpas* clearly show about the dependence of artists in the area Sārnāth prototypes. While the images of Gayā show the influence of the Mathura school of the Gupta period.

Some of the specimens which bear above mentioned features are as follows:

1. A stucco standing image of the Buddha discovered at Nālandā is one of the best representatives of this section.[159] The image wears a foldless drapery in the style of the Gupta sculptures of Sārnāth, covering both the shoulders. The loose end of the outer garment is raised up to the shoulder level, not held at waist level as common both in Mathura and Sārnāth during the Gupta period, and the hem of this garment falling from the left hand, and the broken right hand seems to have been raised in *Varada- mudrā*. The style of the image is mainly based on the ideals of Sārnāth school.The body retains the slender form of Sārnāth Gupta images, but the pose is stiff and the rhythm clearly shows the distinction between the lower torso, upper torso and head in place of the harmonious flowing composition of Sārnāth Gupta images.

2. The next representative of this section is a seated Buddha image.[160] Here the Buddha is shown seated in *Pralambapāda-āsana* with hands held at chest in the preaching gesture. The image is very much similar to the above mentioned image, however, here we notice that the folds of the garments are rendered by incised lines between the legs, rarely noticed in the Guptan art of Sārnāth. Between the eyebrows there is *Ūrṇā*.

3. An image of the Buddha discovered at Bodhgayā is the next representative of this section.[161] The Buddha is shown seated on the

---

159. *Ibid.*, pl. 68.

160. *Ibid.*, pl. 69.

161. *Ibid.*, pl.62.

coils of a serpent, the hands are folded in the lap (*Dhyāna-mudrā*), The foldless drapery covers both the shoulders . The eyes are half closed and in contrast to upper, the lower lip is thick, and earlobes are elongated. The facial expression shows the resemblance with Sārnāth images which belongs to the period, circa mid- fifth century, when the influence of Mathura was still apparent in the art of Sārnāth. Thus, we may say that the image shows the Mathura influence.

4. The next representative of this section is a seated Buddha image from the Magadhan area.[162] The Buddha is shown seated on the pedestal and his right hand is raised in *Abhaya-mudrā*. The drapery covers the left shoulder only. There are double lines just above the ankles for garment ends.The eyes are half-closed. The hair is arranged in the curly manner and the *Uṣniṣa* is clearly visible. The halo is plain with a prominent line at the edge. There are three lines round the neck. On the pedestal an inscription proclaiming the image to be the pious gift of the general (*balādhikṛta*) Malluka.

5. The Sultanganj bronze Buddha image is the next representative of this section.[163] Here the Buddha is shown in standing pose with his right hand in the *Abhaya- mudrā*. The image relates to the ideals of the Guptan Buddha images, but there are some differences from them, for example, the pose of image is not gentle, flowing *Abhanga* stance of the Sārnāth images, rather a somewhat stiffer pose. The face is full and incised lines are used to delineate the eyebrows and the *Ūrṇā* between the eyebrows is clearly visible.

6. A stucco Buddha image from Nālandā seated in *Pralambapādaāsana* is the next specimen of this section (Pl.26) The image is depicted in a niche and is flanked by two fragmentary deities. The transparent foldless drapery covers both the shoulders, the lower portion of which reaches below the knee. The face and the nose of the image are slightly flat, one ear is elongated and the other is missing. The hair is arranged in curly manner with prominent *Uṣniṣa*, there is *Ūrṇā* between the eyebrows, eyes are half-closed and the halo

---

162. *Ibid*, pl. 63.
163. *Ibid.*, pl. 96.

is plain. There is a pillow like thing behind the Buddha's body and the hands are in preaching pose.

SECTION-3 : The images of the Buddha belonging to the period circa 700-800 A.D. reflect significant changes in styles and features. Earlier the images of Bihar were classified into two styles of art, Sārnāth- inspired Nālandā school and Mathura- inspired Bodhgayā art, but now we find that the sculptures belonging to the period circa 700-800 A.D. are intermixed with each other. The next significant change is the sculpture of Nālandā is now no more architectural adornment but free-standing imagery. Some of the stone sculptures are larger than life size. With the rise of the Pālas in power we find large number of bronze sculptures at Nālandā and other adjoining areas.

Some of the specimens which bear above mentioned features are as follows:

1. A small seated Buddha image from the Bodhagayā region, now placed in Candrasālā above the entrance to the Mahābodhi temple is one of the best representatives of this section.[164] The image is seated on a full blown lotus and the right hand is lowered in *Bhumisparśa mudrā*. The face of the image retains very little of the earlier delicacy. The distinctive roll of flesh beneath the navel indicates the remanant of earlier Nālandā image, which shows the influence of Nālandā style on Bodhgayā images.

2. The next representative of this section is a seated image of the Buddha, in *Pralambapāda-āsana*, from Bodhgayā.[165] Stylistically the image is somewhat based on the ideals of Nālandā. But unlike any previous Buddha image from Bodhgayā, there are parallel lines in the garment which indicates the folds of the drapery. The image is heavy set in the Bodhgayā manner but limbs are awkward. The hands of the image are turned in preaching pose, the eyes are half-closed and the earlobes are elongated. The facial expression of the image reflects the ideals of Nālandā style.

---

164. *Ibid.*, pl. 136.
165. *Ibid.*, pl. 139.

3. A gilt bronze image of the Buddha is the next representative of this section.[166] The Buddha is shown in standing pose and his right hand is in the *Varada- mudrā*. He is contemplative looking inwardly. By his left hand, he holds the upper garment which covers the left shoulder only. The Buddha is standing on a lotus which is placed over a high rectangular two tiered pedestal. The weight of the body of the Buddha rests upon the right leg. Three flowers and leaf devices connect the figure with the *Prabhāmaṇḍala* decorated with a beaded and flame designs.

4. A seated image of the Buddha from Rājgir- Nālandā area is another representative of this section.[167] The Buddha is shown seated on a lotus throne. The high pedestal is carved with recumbent deer and worshippers flanking the central wheel. The hands are turned in the preaching pose. There are hook-shaped flames that encircle the halo. The foldless drapery covers both the shoulders. The face retains the gentle cast of the Sārnāth Gupta style.

5. A seated cross-legged Buddha image kept in Nālandā Museum(Pl.27) is another specimen of this section. The foldless transparent drapery covers both the shoulders and the hands are in *Dharmacakramudrā*. The hair is arranged in curly manner with *Uṣnisa*. The ears are elongated, and there is *Trivali* mark round the neck. The halo is partially broken, the remaining portion shows that it is plain and there is a round line-mark on the edge of the halo and the flames are also decorated on both sides of the halo. On the pedestal a pillow like thing is depicted behind the body. Below the pedestal two deers are depicted seating face to face and in betweem *Dharmacakra* is also depicted.

B. *BENGAL* : Stylistically the images of the Buddha from Bengal belonging to the period under review can be divided into two sections:

SECTION-1 : The ideals of the images of the Buddha from Bengal of the Gupta period are mainly based on Sārnāth school. Only a few images of the Buddha of the Gupta period are known from Bengal.

---

166. *Ibid.*, pl. 143.
167. *Ibid.*, pl.181.

Specimens which bear the above mentioned features are as follows:

1. The earliest extant image of the Buddha discovered in Bengal is Biharail in Rajasahi district (Pl.28). The image is shown in standing pose and it has been dated to the period 5th century A.D. The image bears close affinity to the Sārnāth images, that is why it is said that the image was imported from Sārnāth.[168] The material used in the sculpture is buff-coloured sandstone often used for the Gupta sculptures of Sārnāth.

2. At Panna, near ancient sea-port of Tāmralipta, modern Tamluk (Midnapore district), some terracotta sculptures have been discovered belonging to the Gupta period. Among them there is a small terracotta plaque which depicts a seated figure of the Buddha with the hands in preaching pose. There is a brief inscription that appears on both sides of the halo. Stylistically the images has been ascribed to 5th-6th century A.D.[169]

SECTION-2: The sculptures of Bengal belonging to the period circa 700-800A.D. reflect some special pecularities which distinguish them from others and may be called as the refined ideals of the early period, which later on reached its zenith.

Specimen which bears above mentioned features:

1. A standing Buddha image from Bhāsu-Vihāra, Mahāsthāngarh is one of the best representatives of this section(Pl.29)[170] Here the standing Buddha is in *Varada-mudrā*. The figure is very much close to the Biharail Buddha image. The stance of the image and the left hand raised to the shoulder level as well as the disproportionate hands and feet and harsh modelling, most clearly recall its eighth-century Magadhan counterparts. Moreover, this image stands on a double lotus, popular during the eighth century, and is attended by a kneeling devotee, perhaps the donor.

C.ORISSA: A large number of the images have been discovered in different parts of Orissa which are aesthetically well-refined and

---

168. R.C.Majumdar (Ed.), *History of Bengal...*, p. 523, pl. XLVI, 12.

169. F.M.Asher, *Op. Cit.*, pl.39.

170. *Ibid.*, pl. 232.

mainly based on the ideals of Sārnāth school. But there are some special features of these images e.g. the facial expression does not reflect much affinity with the Sārnāth images. Unlike Sārnāth images, here the eyes of some of the Buddha images are open. The nose is comparitively short and sharp and the lips are as thin but the lower one is thick. The hair is arranged on the curly manner but stylistically they are slightly changed to those of the Sārnāth school. The bronze figures of the period are mostly rubbed. There are only a few images of the Buddha before the eighth century A.D. from Orissa.

Some of the spepcimens which bear above mentioned features are as follows:

1. A seated bronze image of the Buddha from Achutrājpur is one of the best representatives belonging to the early period. The hands of the Buddha are turned in *Dhyāna-mudrā*, the hair is curly, earlobes are elongated and the foldless drapery is very much close to the body.[171]

2. A seated bronze Buddha image in *Vajra- parayaṅkāsana* is another representative of the images belonging to the early medieval period.[172] The right hand of the Buddha is turned in *Abhaya- mudrā* while the left hand bearing a part of the *Uttarāsaṅga* is held above the lap. Behind the image there is an oval halo with raised rim of these plain moulding edges by a beaded line.

3. A seated image of the Buddha from Ratnagiri is another representative.[173] The Buddha is seated here in *Bhumisparśamudrā*. The foldless drapery covers the left shoulder only. The left palm and sole of feet bear wheel-marks. The *Ūrṇā* between the eyebrows is clearly visible. The head along with *Uṣṇiṣa* is covered with short spiral curling hair and around the head there is an elongated oval shaped halo. The Buddha is shown seated on the lotus throne.

4. A seated bronze image of the Buddha from Achutrājpur is another representative belonging to the period under review.[174] Here

---

171. D.Mitra, *Bronzes from Achutrajpur*, Orissa, Delhi, 1978,ph.No.36.

172. *Ibid.*, ph.No.38.

173. D.Mitra, *Ratnagiri*, Vol.II, ASI, New Delhi, 1981, pl.CLXVIB.

174. D.Mitra, *Bronzes from Achutrajpur*....ph. No. 40.

also the Buddha is seated in *Bhumisparśa-mudrā*. The foldless drapery covers the right shoulder only. The halo is broken from the upper portion, however, it clearly shows that it has a raised beaded border which is edged by a continuous row of flames and in the central portion, behind the back of the image is open.

CHAPTER - 9
CONCLUSION

The problem of the origin of the Buddha image has not been unanimously resolved. Scholars are divided into three camps. Foucher, Smith and their like minded scholars advocate the Gandhāra origin of the Buddha image. According to them the images of the Buddha first appeared in the North-Western region of ancient India i.e. Gandhāra in and around the christian era due to foreign inspiration who were settled in North-Western region. They used to make the images of their own deities, so there was no problem for them to carve the image of the Buddha and this is why the images from Gandhāra reflect Greek, Roman and other foreign influences. It is said that the Gandhāran Buddha images are somewhat similar to the Greek God *Apollo*. According to them Indians were using only aniconic symbols to represent the Buddha and other deities. But this theory is aptly rejected by A.K.Coomaraswamy, Van Lohuizen, V.S.Agrawala etc. It is quite obvious that image making was already in vogue in India atleast before the emergence of the Buddha image. The sculptors, who can carve the Parkham Yakṣa and the beautiful Didarganj Yakṣi images, would not have any problem to carve the anthropomorphic images of the Buddha. It was the contemporary socio-religious needs which necessitated the carving of the anthropomorphic images of the Buddha.

The history of Buddhist art can be traced back from the days of Aśoka through his monolithic pillars with some symbolic capitals viz., wheel, lion etc. which are supposed to be mythologically related to the Buddha. Though these symbols were equally significant for Brāhmanism but since Aśoka's leaning towards Buddhism and as he had done a lot to the cause of Buddhism we may say confidently that the symbols exhibited on the Aśokan pillars were related to the

Buddha. The monuments of early Buddhism at Bharhut, Sānchi, Bodhgayā and Amarāvati etc. are symbolic and illustrate, *Inter-alia*, the different events of the life of the Buddha and the *Jātaka* stories.

The early Buddhist canonical literature does not mention about the making of the anthropomorphic images of the Buddha, but some later Buddhist texts like *Divyāvadāna*, state that the anthropomorphic image of the Buddha was made even during the life time of the Master. This is also corroborated with the accounts of the Chinese travellers, Fa-Hien and Huien-Tsang. But these sources are not corroborated with the archaeological evidence which shows the later emergence of the Buddha image. However, one thing is notable here that the account of *Divyāvadāna*, Fa-Hien and Huien-Tsang mention that the images of the Buddha which were made during the life time of the Master, were in wood and cloth which can not last such a long period. Any way, it is clear that the early Buddhist art was symbolic.

The aniconic nature of early Buddhist art was in keeping with the contemporary attitude in higher religions in which the deities were conceived aniconically. So, there was no need for image- making and worship. It was only the local folk deities and little tradition which made image of their deities before the christian era. Even in the Theravāda tradition the Buddha was conceived as a human being, and the Buddha identified himself with the Dhamma. Apart from this, the nature of early Buddhism also suggests that there was canonical interdiction to depict the Buddha in human form.

Now the question arises: what necessitated the Buddhists to change their concept from aniconism to anthropomorphism?; precisdy we may say that it was the outcome of gradual sectarian development in Buddhism. We find the references that in Theravāda Buddhism the Buddha was considered as a human being who guided the people for emancipation, but gradually he was attributed with certain superhuman qualities. In Theravāda the material body of the Buddha was considered as *Putikāya* (a body full of impure matters) where as the *Dhammakāya* (collection of Dhammas) was considered as the real body, but it has no form. The Mahāsaṅghikas considered the Buddha as *Lokottara* (supramundane) and they deified the Buddha. The

Sarvāstivādins conceived the Buddha as a human being but at the same time they attributed to him the divine and superdivine powers.

Devotion was in Buddhism from the earliest period but not as a cult. Gradually the Buddha became an object of adoration and meditation. The Mahāsaṅghikas laid special emphasis on the Buddha-*Bhakti* and enunciated that by worshipping the Buddha one can achieve the emancipation. The *Aṣṭasāhasrikā-prajñāpāramitā Sūtra* reflects devotionalism as a medium to gain merit. Thus, we may say that the *kāya* concept, deification of the Buddha and the evolution of the *Bhakti*-cult in Buddhism led to the emergence of the Buddha image. But it is said that it was the Brāhmanic *Bhakti* movement which permeated the idea of devotionalism in Buddhism which was at its peak during the Kuṣāṇa period, however, it can not be accepted that Brāhmanic *Bhakti* movement was solely responsible for the emergence of the anthropomorphic images of the Buddha. Actually, the contribution of Brāhmanic *Bhakti* movement was that it gave impetus to infuse the personified form of devotion in Buddhism but the process of deification of the Buddha was already started by the Mahāsaṅghikas, Andhakas and Sarvāstivādins respectively; the Brāhmanic *Bhakti* movement only gave impetus to this process.

The popularity of Yakṣa and Nāga cult also contributed to the emergence of the Buddha image. The early images of the Buddha/Bodhisattva from Mathura show a great affinity to the images of Yakṣa and Nāga.

The economic prosperity of the period also equally contributed to the emergence of the Buddha image. It is quite obvious that people can invest money to this purpose only in surplus financial condition. The period of first century B.C.-A.D., in which the images of the Buddha are supposed to have been made, for the first time, is considered to be the economically prosperous one. Lastly, we may say that the sectarian development in Buddhism, Brāhmanic *Bhakti* movement, popularity of folk-cult (Yakṣa and Nāga) and material prosperity of the period contributed to the emergence of the Buddha image.

The priority of the place where the first images of the Buddha/Bodhisattva evolved has not been solved unanimously mainly due to the lack of adequate material. In this respect, as stated earlier, scholars are divided into three camps.

1. The theory of the Gandhāra origin of the Buddha image is led by Foucher and his like minded scholars.

2. The theory of the Mathura origin of the Buddha is advocated by A.K.Coomaraswamy, V.S.Agrawala etc. and thirdly, 3. Theory of Udyāna-Kashmir origin of the Buddha image recently put forth by A.K.Narain.

In Gandhāra region the images of the Buddha were known from Kujula Kadaphises period as evidenced by a Kharoṣṭhi inscription of the king Senavarma. The discoveries of a number of sculptured panels, displaying, *inter alia*, the images of the Buddha in a group of monuments discovered at Butkara(Pakistan) have been dated to the period from late 1st century B.C. to early first century A.D. The reliquary of Bimaran is not so early as considered by some scholars. The theory of Foucher that the first image of the Buddha was carved in the Gandhāra region, no longer stands.

The Mathura school also carved the images of the Buddha in the same period i.e. between 1st century B.C. and A.D. but named it Bodhisattva due to canonical interdiction.

The theory of A.K.Narain that the first image of the Buddha/Bodhisattva was made in the Udyāna-Kashmir region, is a new addition, but there are some reservations in accepting this theory.

(1) Whether the region Udyāna-Kashmir can be considered as politically separate from Gandhāra, most of the works mention Kashmir-Udyāna as a part of Gandhāra. Even in early Buddhist literature both Gandhāra and Kashmir are mentioned together.

(2) No doubt the Śakas patronised Buddhism but the coins of Maues, which depicts a cross-legged human figure, can not be identified as the Buddha/Bodhisattva figure with centainty as the figure is very much rubbed. Buddhist image discovered from the Kashmir region mostly belong to the early medieval period. Thus, Narain's hypothesis cannot be considered as conclusive.

Thus, the choice lies in between Gandhāra and Mathura. The stylistic features of both the schools (Mathura and Gandhāra) reveal that both of them developed independently and most probably both Gandhāra and Mathura schools produced the images of the Buddha/Bodhisattva simultaneously in the period circa 1st century B.C.-A.D.

The features of the earliest images of the Buddha/Bodhisattva from Mathura school are not impressive: rather crude and rudimentary and symbols were dominating, though the human form was also emerging but reluctantly. The reluctance is gathered either by a tiny figure of the Buddha or through captioning the represented deity as Bodhisattva and not as the Buddha. This hesitation continued in the early Kuṣāna period also. A close resemblence is noticed between the seated Buddha image and the Jina image on the *Ayagapaṭas* and between the standing figures of the Buddha/Bodhisattva and the early Yakṣa images.

During the Kaniṣka period the shape of the images were not uniform and it differed from image to image. Generally two motifs are noticed : either lion or a cluster of flowers. The lion seems to suggests the supremacy of the Buddha Śākyamuni who was also known as Śākya-siṁha i.e. the lion among the Śākyas. The cluster of flowers generally assuming the shape of the royal turban so it may be identified as the turban of Siddhārtha which was thrown away by him along with his hair during the great event of the renunciation and later it was installed in the heaven by Indra and was worshipped as Cuḍāmani. The sculptors of Mathura has always shown Buddha in *Abhayamudrā* in the early phase of art and other *Mudrās* were introduced gradually later on. The sculptors of Mathura were conceived the Buddha as a superhuman being, a stage which he attained after the enlightenment, that it why the images of the Buddha from Mathura are considered as idealistic in nature. Whereas the Gandhāran sculptors conceived the Buddha as a greatman and consequently they displayed the events of the life of the Buddha and Gandhāran Buddha images are considered as realistic in nature.

The post-Kaniṣka epoch introduced the Gandhāran impact on the images of Buddha at Mathura. The image of the Gupta period became slim and slender and the expression reflected the inner bliss and serenity which clearly manifests the advancement of image-making technology in Mathura school of art.

The images of the Buddha discovered at Butkara I (Pakistan) belonging to the period of first century B.C.-A.D. are considered as the earliest representation of the Buddha in the Gandhāran region. The image of the Buddha in Gandhāra did not inherit the technique of the statue from the early Indian school in particular. The execution of the eyebrows, the eyes, the mouth, the ear and the nose are different from those of Mathura school. But the style of hair and the folds of drapery are similar to the Hellenistic images. The *Ūrṇā*, earlobes, the drapery, the *mudrās* and the way of sitting were according to the Indian ideals or modes. In a sentence we may say that the Gandhāran Buddha images were made with Indian ideas, but not entirely by Indian techniques and partially by Hellenistic technique.

The images of the Buddha appeared at Sārnāth school during later part of the Kuṣāṇa period but those images were supposed to be imported from Mathura. The images of this period are the prototypes of Mathura images. During the Gupta period Sārnāth emerged as an important centre of Buddhist art. The faces of the images are now oval and the hair is arranged in short curly manner. The drapery is foldless and reduced in volume. The halo is generally replaced by a large almond shaped back piece which is larger than the images.

Kashmir school of art yielded a number of Buddhist images in early medieval period. The features of the images reflect the influences of both Gandhāra and Mathura schools. But gradually Kashmir established its own ideals. Here we find some ornamented images of the Buddha. In Kashmir we find sometimes a small figure clearly of Graeco-Roman origin and even the elephant, the horse, the peacock, and the *garuḍa* etc. were also associated with the Buddha images.

The images of the Buddha have been shown in different *mudrās* in Western Indian caves. The features of the images of the Buddha are the synthesis of the ideals of Gandhāra, Mathura and Sārnāth schools. Here the images of the Buddha are both in round and in relief.

The images of the Buddha from Andhra school , are considered as the prototype of the images of the Buddha of the Mathura school. But, here the *Uṣṇīṣa* is low and round and the head is covered with curls of hair. The curls of hair are very similar to those on the head of the Jina images of Mathura belonging to the Kuṣāṇa period. The face is oval and the cheeks are full. The eyebrows are in remarkable relief. The eyes are nearly oval with definite lines at the top and the bottom. The lower-lip is broad, exactly similar to the Mathura ones in the Kuṣāṇa period. The corners of the mouth is slightly turned up. The style of drapery is slightly different from the way in Mathura and it is not so deep as in the case of Gandhāran images. The drapery covers only the left shoulder. The lower end of the drapery turns upward which is a feature of the Andhra school.

The images of the Buddha have been discovered from different parts of Bihar viz., Bodhgayā, Kumrāhār, Nālandā, Sultanganj etc. The styles of the images are mainly based on the synthesis of the ideals of both Mathura and Sārnāth schools. But the images from Nālandā had its own ideals particularly the images which belong to the period circa 700-800 A.D.

The early images of the Buddha from Bengal are mainly based on the ideals of Sārnāth school, however the images from Bengal show some refinement.

The images of the Buddha from Orissa school are partially influenced by both Mathura and Sārnāth schools. The facial expression, though expressing serenity and contemplation, does not reflect much affinity with the Sārnāth images. Here the images are huge and life like.

The whole journey of the images of the Buddha from its inception to the 8th century A.D. manifests diverse facets. The purpose of art in the early period was the manifestation of religious contents, but in the Gupta period, all three aspects, religion, art and aesthetics were harmoniously converged. The Buddha images of the post-Gupta period are generally flanked by Tāntric Buddhist deities. The image of the Buddha from the earliest period to 8th century A.D. also reflect some influence of different sects of Buddhism.

SELECT BIBLIOGRAPHY

Adhya, G.L., *Early Indian Economics*, Asia Publishing House, Bombay.

Adval, N., *The Story of King Udayana*, Varanasi, 1970.

Agrawala, V.S., *Studies in Indian Art*, Vishvavidyalaya Prakāshan, Varanasi, 1965.

----------, *Bhāratiya Kalā*, Varanasi, 1966.

--------, *Gupta Art*, Prithvi Prakāshan, Varanasi, 1977.

--------, *Sarnath*, ASI, 1980.

Ahir, D.C., "Buddhism and Orissa", *Proceedings of International Seminar on Buddhism and Jainism*, Cuttack, 1978.

Aiyappan, A. & P.R. Srinivasan (Ed.), *Story of Buddhism with special reference to South India*, Govt. of Madras, 1956.

Asher, F.M., *The Art of Eastern India*, 300-800, Oxford University Press, 1980

Ashton, L.,(Ed.), *The Art of India and Pakistan*, London, 1947-48.

Bachchofer, L., *Early Indian Sculpture*, Vol.II, Munshiram Manoharlal, Delhi, 1973.

Bailey, H.W., "A Kharoṣṭri Inscription of Senavarma King of Oḍi", *JRASB*, 1980.

Bajpai, K.D., "Bodhisattva Maitrey in Early Art", K.K. Ganguli and S.S. Biswas (Ed.), *Rupanjali* (in memory of O.C. Gangoly), Calcutta, 1986.

----------, "Buddha and Maitreya in Early Indian Art", in Devendra Handa (Ed.), *Recent Studies in Indology*, Vol.II, Sandeep Prakashan, Delhi, 1989.

---------, *Mathura*, 1953.

Bamzai, P.N.K., *A History of Kashmir*, Delhi, 1962.

Banerjee, A.C., *Sarvāstivāda Literature*, Calcutta, 1957.

Banerjee, J.N., *The Development of Hindu Iconography*, Calcutta, 1956.

Banerjee, R.D., *The Age of the Imperial Guptas*, Delhi, 1981.

Bapat, P.V. (Ed.), *2500 Years of Buddhism*, Publication Division, New Delhi, 1976 (reprint).

Barua, B.M., "Hāthigumphā Inscription of Khāravela, *IHQ*, Vol. XIV, 1938.

Basham, A.L., *The Wonder that was India*, Rupa & Co. Calcutta, 1981.

----------, (Ed.), *Papers on the Date of Kaniṣka*, E.J. Brill, Leiden, 1968.

---------, "The Background to the Rise of Buddhism", in A.K. Narain (Ed.), *Studies in History of Buddhism*, New Delhi, 1980.

Basak, R.G., *History of North-Eastern India*, Calcutta, 1934.

Beal, S., *Buddhist Records of the Western World*, Vol.I, London, 1906 (reprint in Delhi,1969).

Bhattacharyya, B., *Indian Buddhist Iconography*, Calcutta, 1968.

Burgess, J., *A Guide to Elura cave Temples*, Hyderabad, 1926, p. 16.

Chattopadhyaya, B., *Kushana State and Indian Society*, Calcutta, 1975.

Chattopadhyaya, D.P. (Ed.), *History of Buddhism by Tāranāth*, (tr.) by Lama Chimpa and Alka Chattopadhyaya, Simla, 1970.

Chattopadhyaya, D.P. (Ed.), *Kushāna Studies in USSR*, Indian Studies - Past and Present, Calcutta, 1970.

Chattopadhyaya, S., *Early History of North India*, Calcutta, 1958.

Chaudhury, B.N., *Buddhist Centres in Ancient India*, Sanskrit College, Calcutta, 1982.

--------, "Date of Buddha-worship", *The Mahabodhi*, Vol.75, No.7, July, 1975.

Coomaraswamy, A.K., *The Origin of the Buddha Image*, Munshiram Manoharlal, New Delhi, 1972.

____________, *History of Indian and Indonesian Art*, Munshiram Manoharlal, New Delhi, 1972.

Cribb, Joe, "A Re-examination of the Buddha Images on the coins of King Kaniṣka:New Light on the Origin of the Buddha image in Gandhara Art". in A.K.Narain(Ed.), *Studies in Buddhist Art of South Asia*, Kanak Publication, New Delhi,1975.

Cunningham, A., *Archaeological Survey of India*, Annual Report, Vol.XI Calcutta, 1880.

Dani, A.& F.Khan, "Kushan Civilization in Pakistan" in *Central Asia in Kushan period*, Vol.I, MOSCOW,1974.

Dani,A.H., *Chilas, the city of Nangā Parvat(Dyamar)*, Islamabad,1983.

Dasgupta, K.K., "Aśokan Art-why and how for Buddhist", *Proceedings of Indian History Congress*, Vol.XXXI,1969.

—————————, "Origin of Buddha Image" in *Studies in Ancient Indian History*, (D.C.Sircar CommemorationVolume), Sundeep Prakashan, Delhi,1988.

Dasgupta K.K. & Others(Ed.),*Comprehensive History of India*, Vol.III,Part II, PPH,Delhi,1982.

Deambi, B.K.,& Kaul, *Corpus of Sarda Inscriptions*, Delhi,1982.

Desai, Devangana,"Social Background of Ancient Indian Terracotta", in D.P.Chattopadhyaya (Ed.), *History and Society*, Calcutta,1978.

Dobbins, K.W.,*The Stupa and Vihara of Kaniṣka I*, The Asiatic Society, Calcutta,1971.

Dutt,N.,& K.D.Bajpai, *Development of Buddhism in Uttar Pradesh*, Lucknow,1956.

Dutt,N., *Mahāyāna Buddhism*, Indological Book House, Delhi,1973.

—————————,*Buddhist Sects in India*,Delhi,1978.

—————————,*Aspects of Mahāyāna Buddhism and its relation to Hīnayāna*, London,1930.

—————————, *Early Monastic Buddhism*, Vol.II, Calcutta,1941.

Eliot, Charles, *Hinduism and Budhism*, Vol.III, London, 1968.

Faccena, D., "Excavations of the Italian Archaeological Mission (ISMEO) in Pakistan:Some Problems of Gandhara Art and Architecture"' *Central Asia in the Kushan Period*, Vol.I, Moscow,1975.

Fergusson J.,*A History of Indian and Eastern Architecture*, Delhi,1967.

Fergusson,J.,Burgess, *Cave Temples of India*, London, 1880.

Foucher,A., *The Beginning of Buddhist Art and other Essays*, Delhi,1972.

Gangoly, O.C., *Andhra Sculpture*, Govt. of Andhra Pradesh, Archaeological Series No. 36, Hyderabad,1973.

___________,"The Buddha Image". *The Mahabodhi*, Vol.61,No.5 & 6.

Ganhar, J.N., "Buddhism in Kashmir." *The Mahabodhi*,
Vol.67,1959,p.148.

Gills, H.A., *The Travels of Fa-Hien*(tr.), Cambridge,1923,(reprint),
London,1966.

Gokhale,B.G., "Buddhism in Gupta Age", in B.L.Smith (Ed.), *Essay
on Gupta Culture*, Delhi,1983.

___________, "The Early Buddhist Elite", *Journal of Indian History*,
1965,Vol.XLII, part II.

Goswamy ,J., *Cultural History of ancient India*, Delhi,1978.

Gupta, S.K., "Causes of the Absence of Buddha image in Early Indian
Art", *K.P.Jaiswal Commemoration Volume*, Patna.

Gupta, S.P.(Ed.), *Kushāna Sculptures from Sanghol*, Vol.I, National
Museum, New Delhi,1985.

Grunwedel, A.,*Buddhist Art in India*, S.Chand & Co., Delhi,1972.

Harle,J.C., *The Art and Architecture of Indian Subcontinent*, Penguin
Books,1986.

________, *Gupta Sculpture*, Oxford, 1974.

Hartel, H., "Some Results of the Excavation of Sonkh-A Preliminary
Report" *German Scholars on india*, Vol.II, Bombay,1976.

Hazra, K.L., *Royal Patronage of Buddhism in Ancient India*, D.K.Publi-
cation, New Delhi,1984.

Hussain F.M., *Buddhist Kashmir*, New Delhi, 1973.

Ingholt, H., *Gandhāra Art in Pakistan*, Pantheon Books, New York,
1957.

Jaini, P.S., "On the Buddha Image", in A.K.Narain(Ed.), *Studies in
Pāli and Buddism*, Delhi,1979.

Jha D.N.& K.M.Shrimali (Ed.), *Prāchina Bhārata kā Itihāsa (Hindi)*, Delhi University, 1986.

Kak, R.C., *Handbook of the Archaeological and Numismatic Sections of the S.P.S.Museum*, Srinagar,Calcutta,1923.

__________,*Ancient Monuments of Kashmir*, Calcutta, 1923.

Kaul, G.L., *Kashmir Through the Ages*, Srinagar,1963.

Konow,Sten, *Kharosthi Inscriptions* Vol.II, Varanasi, 1969.

Kramrisch, Stella, *Indian Sculpture*, Delhi,1981.

Krishan,Y., "Was Gandhara Art a product of Mahāyāna Buddhism?,*Journal of Royal Asiatic Society of Great Britain and Ireland*, 1962.

Lahiri, A.N., *Corpus of Indo-Greek Coins*, Poddar Publication, Calcutta, 1965.

__________, "The Sarvāstivāda,..."(*Proceedings of Annual Seminar on Sarvāstivāda and its trdition*,Deptt. of Buddhist Studies, University of Delhi,1986.

Lahiri, Bela, *Indigenous states of Northern India*, Calcutta, 1974.

Legge, J., *A Record of Buddhistic Kingdom*, New York, 1965.

Lohuizen-de-Leeuw, J.E.V., *The Scythian Period*, Leiden,1949.

Majumdar, A.K., *Concise History of Ancient India*, Vol.I, New Delhi,1977.

Majumdar, N.G., *A Guide to the Sculptures in Indian Museum*, Part II, ASI,Delhi,1937.

Majumdar, R.C., *Ancient India*, Motilal Banarsidas, Delhi,1974.

__________,(Ed.), *The Age of Imperial Unity*, Bombay,1953.

__________, *The Classical Age*, Bombay,1954.

__________, *The Age of Imperial Kanauj*, Bombay, 1955.

__________, *The Struggle for Empire*, Bombay,1979.

__________, *History of Bengal*, Vol.I, University of Dacca, 1963.

Marshall, J., *The Buddhist Art of Gandhāra*, Cambridge,1960.

__________, *Taxila*, Vol.I, Cambridge, 1951.

Mitra,Debala, *Buddhist Monuments*, Sahitya Samsad, Calcutta, 1971.

__________, *Bronzes from Achutrājpur, Orissa*, Agam Kala Prakashan, Delhi,1978.

__________, *Ratnagiri(1958-61), 2Vols.*, ASI,New Delhi,1981.

Nakamura, H., *Indian Budddhism : A Survey with Bibliographical Notes*, KUFS, Publication, Japan, 1980.

Narain A.K., *The Indo-Greeks*, Oxford University Press, New Delhi, 1980(reprint).

__________, "First Images of the Buddha and Bodhisattvas: Ideology and Chronology" in *Studies in Buddhist art of South Asia*, New Delhi,1985.

Narain, J.& R.Tiwary, "Lalitaditya Muktāpiḍa, The first Antiquarian King of India" in K.K.Dasgupta & Other (Ed.), *Studies in ancient Indian History*, Delhi,1988.

Naudou, J.N., *Buddhists of Kashmir*, New Delhi, 1980

Nehru, Lolita, *Origins of Gandhāran Style*, Oxford University Press, New Delhi, 1989.

Nilakanta Shastri, K.A., *Dakshina Bhārata Kā Itihāsa,*(Hindi) Patna, 1986.

Pal, P., *Bronzes of Kashmir*, New Delhi,1975.

Parimoo, R., *Life of Buddha in Indian Sculpture*, Kanak Publication, New Delhi,1982.

Prasad, B.R., *Art of South India-Andhra Pradesh*, Sundeep Prakashan, Delhi,1980.

Przyluski, J., *Legends of the Emperor Aśoka in Indian and Chinese Texts*(tr.) by D.K.Biswas, Calcutta, 1967.

Ramchandran, T.N., *Buddhist Sculptures from a Stūpa near Goli Village*, ASI, Madras,1929.

Ray, N.R., *Idea and Image in Indian Art*, Delhi,1973.

Ray.S.C., *Early History and Culture of Kashmir*, Calcutta,1957.

Raychaudhury, H.C., *Political History of Ancient India*, Calcutta University, 1972 (reprint).

Rowland B.,, *The Art and Architecture of India*, Penguin Books, 1971.

Sahni, D.R., *Archaeological Survey of India Report*, 1908- 09.

Sahu, N.K., *Buddhism in Orissa*, Utkal University, 1958.

Saraswati, S.K. *A Survey of Indian Sculpture*, Calcutta, 1957.

Sarkar H., & S.P.Nainar, *Amarāvati*, ASI, New Delhi.1972.

Sarkar H.,& B.N.Mishra, *Nāgārjunakoṇḍa*, ASI, New Delhi, 1980.

Sastri P.S., "The rise and growth of Buddhism in Andhra, *IHQ*, Vol.XXXI,1955.

Sengupta, Sudha, *Buddhism in the Classical Age*, Sundeep Prakashan, Delhi,1985.

__________, "Buddhism in Western India", *The Mahabodhi*, Vol.63, Calcutta.

Senmajumdar G., *Buddhism in Ancient Bengal*, Calcutta, 1983.

Sharma, R.C., *Buddhist Art of Mathura*, Agam Kala Prakashan, Delhi, 1984.

Sharma, R.S., *Perspective in Social and Economic History of Early India*, New Delhi,1983.

Sircar, D.C., *Select Inscriptions*, Vol.I, Calcutta, 1942.

__________, *Select Inscriptions*Vol.II, Delhi,1983.

__________, "Eastern India and the Kushans", in *Central Asia in the Kushan period*, Vol.II, Moscow, 1968.

Smith, V.A., (Ed.), *The Cambridge History of India*, Vol.I, Cambridge, 1922.

__________, *A History of Fine Art in India and Ceylon*, Bombay, 1969.

Srivastava, B.B. *Prāchina Bhāratiya Pratimā Vijñāna evam Murtikalā* (Hindi), Sikshā Prakāshan, Varanasi, 1981.

Snellgrove, D.L.(Ed.), The Image of the Buddha, Vikash publishing House, New Delhi,1978.

Stein, M.A. (Ed.), *Rājatarangini*, (Tr.), Varanasi, 1961.

Subramanian, K.R. *Buddhist Remains in South India and Early Andhra History*, New Delhi,1981.

Tarn, W.W., *The Greeks in Bactria and India*, Delhi,1980.

Thaper, Romila, *Aśoka and the decline of the Maurya*, New Delhi, 1980, (reprint).

Vogel, J., *Buddhist Art in India, Ceylon, and Java*, Delhi,1977

————, *Catalogue of the Archaeological Museum at Mathura*, Allahabad, 1910.

Watters, T., *On Yuan Chwang's Travel in India*, Vols.I, II, London, 1904-05.

Wauchop, S.R., *The Buddhist Cave Temples of India*, New Delhi,1981.

Weiner, S.L., *Ajanta, its place in Buddhist Art*, California, 1977.

Williams, J.G. *The Art of Gupta India-Empire and Province*, Heritage Publishers, New Delhi,1983.

Winternitz, M. *History of Indian Literature*, Vol.II, New Delhi,1977.

G L O S S A R Y

Abhaṅga : A standing posture in which the figure displays a slight flexion as a result of the emphasis of the weight of the body on one leg.

Abhaya mudrā : The hand gesture of fearlessness or reassurance. The right hand is lifted at a level of the right shoulder while the palm of the hand, with the fingers extended upwards, faces the onlooker.

Añjali mudrā : In this hand-pose, the two hands are joined palm to palm against the chest, indicating respect, salutation or adoration.

Āsana : The sitting attitude.

Bhadrāsana : A seated posture in which both the legs hang down vertically from the seat. Also called *Pralambapāda*.

Bhūmisparśa mudrā : A hand gesture in which the hand with the palm turned inward and the fingers extended downward touches the earth. The gesture symbolizing his invocation of the earth-goddess as witness of his resistence to Māra.

Cakra : The wheel, sybolizes the Buddhist Law or Buddhist *Dharma*.

Cāmara : The fly-whisk or chauri, symbolic for compassion.

Dharmacakramudrā : The gesture of preaching or turning the wheel of the law, usually associated with the Buddha's first sermon at the Deer park, Sārnāth. The right hand is held near the chest with the united tips of the thumb and index, touching one of the fingers with the palm inwards.

Dhyānamudrā : The gesture of meditation. The hands rest in the lap, the right on the left with all fingers extended, and the palms turned upward.

Dvārapāla : Door guardian.

Jatāmukuta : An elaborate dressing of matted hair in the shape of a crown.

Lakṣanas : In Buddhism, the thirty-two auspicious signs of the Buddha.

Mudrā : Hand gestures with symbolic meaning.

Padma : Lotus.

Padmāsana : A seated posture in which the two legs are folded towards the abdomen, the soles upturned. Synonymous with *Vajraparyaṅkaāsana*. The word is also used to designate the lotus seat.

Pralambapādaāsana : See Bhadrāsana.

Sanghāti : The monastic robe worn by the Buddha or any other monk.

Ūrṇā : One of the thirty-two *Mahāpuruṣalakṣana*, a protuberance on the forehead between the eyebrows of the Buddha.

Uṣniṣa : One of the thirty-two *Mahāpuruṣalakṣana*, a protuberance on top of the head of the Buddha.

Vajraparyaṅkaāsana : A seated posture in which the legs are firmly locked with both the soles turned upwards. Synonymous with *Padmāsana*.

Vitarkamudrā : The hand gesture of argumentation. The index joins the thumb forming a ring while the other fingers remain extended.

INDEX

K

Pl. 1A :   Parkham Yakṣa, Parkham(Mathura),(Photo-self)

Pl. 1B · Seated Buddha, Ahichchatra(Mathura) (Photo courtesy National Museum, New Delhi)

Pl. 2   Seated Buddha, Mathura, 2nd century A.D.,(Photo courtesy : National Museum, New Delhi).

Pl. 3 : Standing Buddha, Mathura, 5th century A.D., (Photo courtesy : National Museum, New Delhi).

Pl. 4 :   Seated Buddha, Gandhāra, 2nd century A.D. (Photo courtesy : National Museum, New Delhi).

Pl. 5 :   Seated Buddha, Gandhāra, 2nd century A.D., (Photo courtesy : Archaeological Survey of India, New Delhi)

Pl. 6 : Standing Buddha, Gandhāra, 2nd-3rd century A.D.,
(Photo courtesy : National Museum, New Delhi).

Pl. 7 :   Standing Buddha, Gandhāra, 2nd-3rd century A.D.,
(Photo courtesy : National Museum, New Delhi)

Pl. 8 :   Pilaster showing Buddha standing on each side, 2nd-3rd century A.D. Gandhāra, (Photo courtesy : National Museum).

Pl. 9 : Standing Buddha, Gandhāra, 2nd-3rd century A.D.
(Photo courtesy : National Museum, New Delhi).

Pl. 10 :   **Seated Buddha, Gandhāra, 3rd century A.D.**
(Photo courtsey) : National Museum, New Delhi).

Pl. 11 :   Standing Buddha, Sārnāth, 5th century A.D.
(Photo courtesy : National Museum, New Delhi).

Pl. 12 :   Standing Buddha, Sārnāth, 5th century A.D.
(Photo courtesy : National Museum, New Delhi).

Pl. 13 :   Standing Buddha, Sārnāth, 5th century A.D.
(Photo courtesy : **National Museum, New Delhi**).

Pl. 14 · Seated Buddha, Sārnāth, 6th century A.D.,
(Photo courtesy : National Museum, New Delhi).

Pl. 15   Seated Buddha, Kahsmir, Srinagar Museum,
(Photo courtesy : Archaeological Survey of India, New
Delhi).

Pl. 16 :   Standing Buddha, Kashmir, Srinagar Museum,
(Photo courtesy : Archaeological Survey of India, New
Delhi).

Pl. 17 : Standing Buddha (Bronze), Kashmir, Srinagar Museum, (Photo courtesy : Archaeological Survey of India, New Delhi).

Pl. 18 :   Seated Buddha, Ajanta, Cave No. 2, (Photo courtesy : Archaeological Survey of India, New Delhi).

Pl. 19   Standing Buddha, Ajanta, Cave No. 19, (Photo courtesy
Archaeological Survey of India New Delhi)

Pl. 20   Seated Buddha, Ellora, Cave No. 10, (Photo courtesy Archaeological Survey of India, New Delhi).

Pl. **21** :   Standing Buddha, Bagh, Cave No. 2, (Photo courtesy :
Archaeological Survey of India, New Delhi).

Pl. 22 : Seated Buddha, Nāgārjunakoṇḍa, (Photo courtesy Archaeological Survey of India, New Delhi).

Pl. 23 :   Seated Buddha, Nāgārjunakoṇḍa, (Photo courtesy :
Archaeological Survey of India, New Delhi).

Pl. 24 :   Standing Buddhas, Amarāvatī, (Photo courtesy : Archaeological Survey of India, New Delhi).

Pl. 25 :   Standing Buddha, Bezwada Museum (Photo courtesy : Archaeological Survey of India, New Delhi).

Pl. 26 :   Seated Buddha (Stucco), Nālandā, 'Photo courtesy
Archaeological Survey of India)

Pl. 27 : Seated Buddha, Nālandā Museum, (Photo courtesy :
Archaeologcial Survey of India).

Pl. 28   Standing Buddha, Biharail, (Photo courtesy   Asher *The Art of Eastern India*, Plate No.36).

Pl. 29 :   Standing Buddha, Bhāsu-Vihār, Mahasthanagarh, (Photo courtesy : Asher, *Op.Cit..* Plate No.232).

Pl. 30 :   Seated Buddha, Ratnagiri. (Orissa)